RISK, MANAGEMENT AND PROCUREMENT IN CONSTRUCTION

Edited by

John Uff
Nash Professor of Engineering Law
King's College London

A. Martin Odams
Research Assistant
King's College London

Centre of Construction Law
and Management
King's College, London

1995

Published in Great Britain by
Centre of Construction Law and Management,
King's College London,
The Old Watch House,
Strand,
London WC2R 2LS
44 (0) 171 873 24 46
44 (0) 171 872 02 10

© Centre of Construction Law and Management and
Contributors

ISBN 0-9514866-4-0

A copy of the CIP entry for this book is available from the
British Library.

Printed in Great Britain by the Ipswich Book Company,
Ipswich, Suffolk.

Contents

Part I
Aspects of Risk

Part II
Management and Procurement

Part III
Particular Issues

Part IV
Choice of Contract

Preface

This volume continues the series of publications of the Centre of Construction Law and Management at King's College London based on proceedings of the Annual Conferences, the present being the Seventh Annual Conference of September 1994. This was organised in conjunction with the Construction Industry Research and Information Association (CIRIA) and we are grateful to Dr. Peter Bransby the Director-General, who chaired the conference, and to a number of contributors whose papers were based on research carried out under CIRIA's research programme.

The conference itself was a timely opportunity to review progress on the recently published Latham Report: *Constructing the Team*. We were particularly pleased that Sir Michael Latham was able to be the Keynote Speaker, giving the conference a particular object and direction. Sir Michael's Keynote Address is included at the front of this volume.

In selecting papers for publication we have had in mind other and wider developments in the areas of risk, management and procurement and we have taken the opportunity to include a significant number of contributions from other sources. Part I of this volume *Aspects of Risk* includes an updated version of John Scriven's paper *A Funder's View of Risk in Construction*, first produced for the Fifth Annual Conference of CCLM. In Part II *Management and Procurement*, we have included a paper by Dr. Michael O'Reilly *Risk, Construction Contracts and Construction Disputes*, which developed from an earlier published paper and from a lecture delivered by Dr. O'Reilly at the Centre. This section also includes a major paper *Impact of Risk Allocation and Equity in Construction* by Ashley, Dunlop and Parker originally produced for, and published with kind permission of, the University of Texas at Austin. Their work has been known to the editors for some time in earlier versions and we are pleased to have the opportunity of including this paper, which is not otherwise readily accessible to UK readers.

In Part III, *Particular Issues*, we are pleased to include a paper *Ground Uncertainty Effects on Project Finance* by Ian Whyte, which is a further development of research work which is referred to in Paper 3 of this volume. Mr. Whyte's analysis of uncertainty in terms of mathematical and financial risk forms a compelling contrast to the necessarily generalised analysis in terms of legal responsibility. Part III also includes, as two separate chapters, the papers on *Major Infrastructure Projects and their Environmental Impact* by Helen Payne and Joanna Higgins, both former students of the Centre of Construction Law and Management 1991-93, these papers being based on their final year dissertations. Joanna Higgins was awarded the Society of Construction Law Prize for the best dissertation and Helen Payne the SCL Hudson Prize for the best paper submitted in 1994. A version of these papers was delivered as the CCLM lecture to the Society of Construction Law in December 1994. This volume provides a timely opportunity to give wider circulation to these important papers.

Finally, in Part IV, *Choice of Contract,* in addition to the papers delivered at the Conference, we have included two additions. *Structuring Contracts for the Achievement of Effective Management* by John Perry reproduces the invitation lecture delivered by Professor Perry in the Michael Brown Foundation lectures in 1993. It was apposite that the subject of this lecture was the research programme underlying development of the New

Engineering Contract, which forms a substratum to much of the debate in and surrounding the Latham Report. We were also pleased to have the opportunity to publish Max Abrahamson's new paper *Risk, Management, Procurement and CCS*. The research and development which led to Professor Abrahamson's new Construction Contract System goes back to his paper delivered at the CCLM's first Annual Conference in 1988 which was subsequently developed through the NEC Drafting Team. As the paper records, the CCS subsequently acquired a life of its own. The CCS is now available as a succinct and more radical alternative to NEC, and it is clearly right that this should form an integral part of the current debate on forms of contract and procurement.

The Editors express their gratitude to all those who contributed to the Seventh Annual Conference and have undertaken the subsequent work of updating and revising their papers, and to the contributors of the additional papers included in this volume, whose work has added significantly to the coverage which we have been able to achieve in this volume. We are also greatly indebted to the staff at the Centre of Construction Law and Management, Mrs. Pauline Gale, Manager, and Mrs. Helen Brownstone, her assistant, for their tireless efforts in organising and collating the material for this volume.

John Uff QC
Martin Odams
King's College London
May 1995

Contributors

Max Abrahamson, BA, LLB, FCIArb.Consultant with Baker & McKenzie, London; Visiting Professor, Centre of Construction Law

John Barber, MA, LLB, CEng, MICE, FCIArb. Michael Brown Lecturer at the Centre of Construction Law.

Gordon Bateman, CEng, FCIArb. Manager of Procurement, Engineering Division, Thames Water plc

Professor Phillip Capper MA (Oxon), BA (Dunelm). Solicitor, Partner with Masons, London. Masons Professor of Law at the Centre of Construction Law.

Dr. John Connaughton, Partner with Davis Langdon Consultancy

Richard Dobson, FRICS, ACIArb. Partner with Dobson White Boulcott.

Gary France, FFB, Director, MACE

Patrick Godfrey, BSc, MICE, Senior Executive with Sir William Halcrow & Partners

Joanna Higgins, BA, MA(Cantab) MSc. Solicitor with Ashurst Morris Crisp

Dr. Will Hughes. Lecturer in Construction Law, University of Reading.

Fin Jardine. Research Manager for general engineering at CIRIA

Dr. Mike O'Reilly, BEng, PhD, LLB, CEng, MICE, ACIArb. Barrister. Lecturer in the Department of Civil & Structural Engineering, University of Sheffield

Helen Payne, LLB(Hons), MSc, Barrister

Professor John G. Perry, BEng, MEng, PhD, CEng, MICE, MAPM. Beale Professor and Head of School, School of Civil Engineering, University of Birmingham.

Malcolm Potter, AA Dipl.RIBA. Consultant Architect with CIRIA

David Richmond-Coggan, MA,C.Eng., MICE, MAPM, Mouchel Management Ltd.

John Scriven, BA. Solicitor, Partner with Allen & Overy.

Professor John Uff QC, PhD, FICE, FCIArb. Nash Professor of Engineering Law at King's College London; Director of the Centre of Construction Law.

Ian Whyte, Eur.Ing., BScTech, Dip.A.S.E., C.Eng, MICE, MIHT, Senior Lecturer, Department of Civil & Structural Engineering, UMIST

Table of Cases

Table of Legislation

UK (Contd)

Table of Statutory Instruments

Table of EC Legislation

Table of Standard Forms

Institution of Civil Engineers

Joint Contracts Tribunal

Keynote Address by Sir Michael Latham

I would first like to pay tribute to the legal profession, including the Centre of Construction Law & Management, and to the Construction Industry Research and Innformation Association (CIRIA), for the very helpful parts which they have played in the work leading to the Latham Review. [1] I received expert guidance from a number of immensely experienced and capable construction lawyers. At least one of these is contributing to this Conference. CIRIA, kindly supplied me with a great deal of information, some of it as yet unpublished, which proved invaluable to the work.

Risk management was one of the key reasons for setting up the Review. Risk is of course endemic in all forms of construction work. It can be managed, minimised, shared, transferred or accepted. It must not be ignored. It is for the client, operating with the appropriate expert advice, whether in-house or specially retained, to decide how much risk is to be accepted as part of the initial project strategy, which will lead in turn to a contract strategy and a procurement route. I tried to explore these important issues in the earlier part of the final report.

Mr. Hugh Clamp and the Central Unit on Procurement provided me with tables[2] which were particularly useful as graphic illustrations of Risk, under the various contractual routes which could be taken. Naturally it would be possible to construct more complex matrices, of which more will be found in the papers which follow.

---

[1] *Constructing the Team, Final Report of the Government / Industry Review of Procurement and Contractual Arrangements in the Construction Industry*, July 1994, HMSO, London.

[2] *Ibid.*, pp 15 -16

The aspirations and needs of clients are seen as being at the core of the construction process, in both the Interim and the Final Reports. This included the stated objective to achieve a cost reduction of 30%. This is, I believe, now widely accepted, because it is manifest commercial sense and clients are insisting upon it. Clients, the construction industry and related professionals have little doubt that this target is feasible. To achieve it, however, demands procurement methods which meet the clients'requirements, as well as team work, contract conditions which reward a win-win approach, effective dispute resolution procedures, greater standardisation of qualification and other paperwork procedures, and a move towards standardisation of components. Several large clients including BAA and Shell thought the target too modest. Others such as Nuclear Electric were looking for further reductions on a basis of partnering.

My Report accepts that if inappropriate risks are transferred through commercial power, or if a procurement route is adopted which cannot cope with the risks, any project is likely to be doomed from the start. Such a project is likely to be claims-ridden and the client will usually be found to be at fault. I therefore tried to place emphasis on the need for proper project and contract strategies and choice of the right procurement route to meet the clients objectives. Particular needs include effective briefing on an iterative basis in order to use to the full the skill of the design team; the integration of the consultants and the contractor's or sub-contractors' design inputs; use of co-ordinated project information or preferably more modern techniques such as knowledge-based engineering; and the compilation of a nexus of contractual documents available to the client covering the whole span of the design and construction process.

I have tried to sound two warning notes about the dangers of less well coordinated documentation or use of inappropriate procurement routes. One was to repeat the wise words of Mr. James Nisbet[3] about the potential fuzzy edges between designer participation; the other was to repeat the sad story of the new glasshouse in Hampshire.[4] It is therefore, particularly encouraging that the authors of the New Engineering Contract have now published their contract documents for professional services and for adjudicators. This will now offer a full matrix of integrated contract documentation in the second edition which is shortly to be published.

When the Review was set up it was specifically designated as a Joint Government Industry Review which also involved clients, not merely a government Review of the Industry. The different participants all had different priorities for the Review. It is no secret that the drafting of the terms of reference proved both lengthy and difficult. In fact the period taken to draft the terms of reference was longer than the period between the appointment of the Reviewer and publication of the interim report, and longer than that between publication of the interim report and the final report going to press.

The primary concern of the sub-contractors, as now organised under the banner of Constructors' Liaison Group but then under other groupings, was the nature of the contract conditions being offered to them by main contractors. The main contractors themselves also complained about the conditions, or at any rate about the amendments or deletions to the standard forms which they were offered by clients. In their turn the clients were clearly unhappy with the standard forms, and this is evidenced by the clients either changing them or producing their own. This observation includes public sector clients. For example, the Department of Transport, now the Highways Agency, devised its own form of contract as an alternative to ICE 5th edition, rather

---

[3]   *Ibid.,* p 24

[4]   *Ibid.* p 26, as graphically described by the National Audit Office and the Public Accounts Committee at footnote 17.

than use the 6th edition. Similarly, the Scottish Roads Directorate has produced its alternative tendering initiative, and the Ministry of Defence has amended GC/Works 1 Edition 3, which was itself seen as the Government's own contract.

That being the trend, I felt that it would be as effective for me to re-offer Banwell's advice of 1964 to use standard forms unamended as it was for King Charles I to tell the Court of Parliament in 1649 that they had not the legal powers to try him. Both statements might be seen as accurate but nobody would have taken any notice and the King still lost his head. I therefore felt that I had to try to think the problem through from the start and to draw up principles by which a contract might be based on team-work, on 'win-win' solutions and on a partnering approach. The objective which I had in mind was to serve the interests of clients, but with fairness to all concerned, hence the principles which I have tried to spell out in the Review itself. While the NEC meets many of these principles, and I hope will meet more of them in its second edition, I did not start out by using it as the benchmark. The Report also recognises that the remaining standard contract forms will continue to be available, amended, as I hope, to meet the principles which I have set out.

I have proposed a Construction Contracts Bill which is intended to support good contractual practice, but not to do what some people urged, namely to enshrine the mandatory use of existing standard forms unamended. Those forms are themselves controversial and many large contractors or first-line specialist contractors use their own in-house forms of domestic sub-contract because they regard DOM/1, or the Blue Form, as inadequate for their purposes.

The Report has generated much activity. It recommended the creation of a Standing Strategic Group for the construction industry involving clients. The Secretary of State has now agreed to chair this group. The Report further proposed the creation of a Construction Clients' Forum with representation of private sector clients, and also public sector if they so wished. That body was officially launched on August 31, 1994; already a number of major players have joined it including the British Property Federation, the CIPS, The Construction Round Table and several others. The CBI has indicated strong support to the Report and their Director General Howard Davis has also stated that his organisation is keen to join in the CCF. This will represent a significant force for change. The Government have also announced that their own procurement practice is to be reviewed to ensure that they adopt the best practices recommended by the Report. The Report proposed an Implementation Forum, with a number of specific duties. This Forum has also been set up and, indeed, met for the first time on September 5th. I agreed to be the Chairman, not without hesitation, because I thought that perhaps a fresh mind would be better. Others thought differently and I felt it churlish to opt out. The Forum has two representatives each from the DoE, the Construction Clients' Forum, the Construction Industry Council, the Construction Industry Employers Council, and the Constructors Liaison Group. Its terms of reference require it first to set up and manage until the end of 1994, through its constituent bodies, a network of Task-Forces constituted to work on implementing the reports recommendations. Those Task-Forces are already coming into being. Eleven have been approved, with chairmen and secretaries appointed and with clear tasks and timescale for their work.

Additionally we are charged with advising the Standing Strategic Group, and thereby Ministers, by the end of 1994 on a number of issues.

- First, which recommendations can be taken forward by existing agencies without any additional structures
- Secondly, how others should be taken forward including legislative implications.

- Thirdly, the timescales and mechanism for the production of the Construction Strategy Code of Practice.
- Fourthly, whether to appoint an Ombudsman to examine allegations of poor practice or issue public statements about them, a matter to which some of the participants in the Review are particularly attracted.
- Fifthly, what is to happen to the Forum after the end of 1994; in particular, whether we should follow the example of Australia and Singapore and set up a Construction Industry Development Agency with a clear life-span to carry the proposals further.

Some recommendations in the Report, including those on contracts, are addressed to other bodies. The ICE, authors of the New Engineering Contract, have already accepted and largely implemented my proposed amendments. The proposal which I made for the JCT and the CCSJC, which reflected suggestions made to me in preparing the Review, are currently being considered by those bodies and their participants. While they involve obvious difficulties, even painful changes, some major participants in both bodies want to go considerably further and faster than I have deemed prudent or possible. If better or speedier ways could be found to implement the principles of what I and others have suggested, I will be the first to welcome them.

There is no monopoly on the proposals to the Report and absolutely no closed season for new ideas and initiatives. Indeed, it is my constant search for such expert feedback and constructive suggestions that brings me to this Conference. I have accepted every invitation which I have received, except those where it was physically impossible to do so. I regard this level of feedback to be essential to the work, just as I am also involved in almost daily one-to-one contact with significant players from all sides of the industry who want to talk to me or whose wisdom I wish to absorb.

The Review was announced by the Minister in July 1993. Four or five days later I was asked to undertake it, with terms of reference which I had not drafted, a timescale which I had not chosen, and with a structure I was given. Several old and wise heads told me that it was impossible, or even if possible, doomed to failure. When the Interim Report appeared, I was told that was the easy part; the tough job would be detailed recommendations. When those appeared in the Final Report, I was told that implementation was always the hardest part. The truth is that none of it has been easy for any of us. If it had been easy there would never have been a Review at all, but it is the wisdom and experience of people such as those presenting papers here who have driven the process forward to this stage. I am sure that we can and will drive it further to construct the team.

Part I

Aspects of Risk

1. Overview of Risk in Construction*

Phillip Capper

Synopsis

This paper examines risk management in the context of contract documents including standard forms. The selection and putting in place a contract framework appropriate to the particular risks is identified as significant to the success and use of project management techniques.

Introduction

The opportunities for contractual risk management are well summarised by Thompson and Perry in their 1992 SERC report.[1]

"Risk management can involve:

- identifying preventive measures to avoid a risk or to reduce its effects

---

* The ideas in this paper have been developed from Professor Capper's 1992 inaugural Lloyd's Register lecture to the Royal Academy of Engineering on the Practical Management of Legal Risks. They are being taken forward in the context of a current Research Project for CIRIA, leading to the production of *"A Client's Guide to Risk Management in the Construction Industry"*. The Guide is being prepared by Sir William Halcrow and Partners Ltd, in association with Laing Technology Group Ltd, Professor Peter Thompson - AMEC Professor of Engineering Management UMIST, and Professor Capper, Masons Professor of Construction Law, King's College, London.

[1] *Engineering Construction Risks. A guide to project risk analysis and risk analysis.* P.A.Thompson & J.G.Perry (Eds.), Thomas Telford, London, 1992.

- proceeding with a project stage-by-stage, initiating further investigation to reduce uncertainty through better information
- considering risk transfer in contract strategy, with attention to the motivational effects and the control of risk allocation
- considering risk transfer to insurers
- setting and managing risk allowances in cost estimates, programmes and specifications
- establishing contingency plans to deal with risks if they occur.

Risk management will not **remove** all risk from projects; its principal aim is to ensure that risks are **managed most efficiently**. The client and his project manager must recognise that certain risks will remain to be carried by the client. This 'residual risk' must be allowed for in the client's estimate of time and cost."[2]

Contractual documents are tools for managing risks. Their purpose is to determine the consequences of particular risks which you must previously have identified. You must have the right tool for the job. This means choosing a contractual framework, for the construction phase, which is appropriate to the objectives and constraints of that project. The aim of this process of choosing an appropriate contract framework is to make more certain the practical and financial consequences of hazards.

An important consideration here is whether risks in construction are to be considered in isolation, or within the financial context of the "whole-life" of the project. The appropriateness of the contract as a risk management tool depends on the nature of your business as client and the experience of the your

---

2   *Ibid.* p.9; and see further extracts in Annexe 1

organisation in relation to the underlying technology of the project.

Contracts are adaptable tools

Contracts are therefore to be viewed as adaptable tools, and not fixed datum points or ends in themselves. In essence the contract has to define the objectives of the project, qualified by its constraints. It does so by defining what each party has to do, and making provision for a range of "what if" scenarios. But, we must recognise that ultimately competence is more important than procedures; that risk assessment serves judgement; and, judgement is practicable only if trust exists between the parties. These latter parameters are also vital for the management of risk, but are beyond the scope of contractual provision. The vital parameters requiring definition in the construction contract are:

- what the client has to do

- what the contractor has to do

- by what dates these various tasks each have to carried out

- pre-defined mechanisms for payment.

The emphasis must be on using the contract documents to reduce uncertainty or provide mechanisms for action when particular uncertainties eventuate into reality. Those mechanisms must in themselves be as clear as possible and based upon defined reference points. It is not an answer to postpone these issues for later judgmental decisions by a contract administrator. Nor should dispute resolution clauses be regarded at the outset as a mechanism for determining the price payable or for fixing in retrospect a reasonable programme. Of course, well planned dispute resolution procedures as part of the construction process can be a way of testing the outcome of some uncertainties.

Are disputes more likely because of traditional contract forms?

It is an error to believe that disputes are necessarily likely to arise in construction. Paradoxically, disputes are perhaps more likely to arise through the use of traditional UK standard forms. As Thompson & Perry found "existing models do not greatly help the achievements of objectives i.e. they do not make a significant contribution to reducing the effects of risk."[3] The contractual frameworks enabling modern construction are changing rapidly. The advance of high technology demands integration into construction projects, across disciplines well beyond those of the traditional construction supervisory team - architect or engineer.

Despite the many interfaces, the construction project as a whole has to be managed within a single coherent legal and contractual framework. Traditional multi-party involvement may yield to Partnering techniques but demand is high for single point responsibility. Responsibility is a legal concept. But it is workable only if it is realistic. For example, turnkey lump sum contracting with employer approval of contractor design may seem very attractive at first sight to the financing community. The reality is that it is not likely to prove as certain as to end result as would more appropriate contractual mechanisms. If a project is let on unduly onerous terms, or it is administered by one discipline whereas substantial components of the project involve other disciplines, there will arise "pinch points". It is the pinch points that provide the fertile ground for growth of legal risks into expensive realities of cost overrun for one party or another.

---

[3]   *Ibid.* p. 36.

Designing the contractual framework: provisions for uncertainty

The most important element in the practical management of legal risk is the design and construction of the contractual framework. Of course words and phrases of contracts are vitally important, but many of the legal risks that require management arise outside the legal words and phrases. They arise through the chosen methods of administration of the contract and through traditional approaches - some now unfortunately out-moded - to be resolution of problems as they arise. The essential reason for having a contract at all is this: to try, so far as legal provisions reasonably can, to render more certain the practical and financial consequences of matters which are physically and factually uncertain at the time of entering into the contract.

It is a very peculiar characteristic of construction projects, that they very often proceed despite high degrees of uncertainty. The participants show remarkable commercial willingness to go forward with uncertainty as to design solutions, uncertainty as to the eventual scope of the works, and uncertainty as to the time periods that will realistically be required in order to complete the works. Further uncertainties necessarily cascade from those, such as out-turn costs, scope of sub-contract works and terms. The norm has been traditionally to postpone addressing uncertainties, leaving enormous discretion to be exercised in the future by those administering the contract. The management of construction risks will be better achieved by more pro-active contractual strategies.

Casting the provisions in context

The ability for all concerned to attribute realistic financial figures, and hence cover for risk, to those uncertainties depend not just upon the particular contractual words or phrases. Rather, it depends critically upon the area of construction and management practice within which it has been chosen to operate the contract, and whether one can achieve a harmony and synthesis between those who are performing the works and those who are procuring and monitoring the works. Nowadays, that is a much more complex grouping. The demands and constraints of those financing and those eventually operating the works may be as significant or more significant than those of the design/construction team. Yet, you have to guard against differing perceptions of particular risks. As Thompson and Perry observed:

> *"If recognised, a risk also tends to be seen differently by engineering, financial, commercial and legal departments, by the eventual users of the completed facility, and by general managers and their advisers. The attitudes, experience and careers of individuals may cause genuine differences of understanding of objectives, and so affect their evaluations and perceptions of risk. Differences can be particularly acute between client, consultants, bankers and insurers on export projects. These 'institutional' risks can seriously affect the initial identification and subsequent management of potential risks crucial to project success."* [4]

---

[4]   *Ibid.* p. 14.

Inappropriateness in the use of traditional forms

There is another aspect of inappropriateness that can be illustrated by reference to the FIDIC and ICE forms. They were of course designed for civil engineering works. Civil engineering works are conventionally treated as relatively uncertain in their scope. What ground conditions will be encountered? What quantities will actually be required? So of course one has to allow for change and resolution of the problems. That is why the FIDIC and ICE forms are remeasurement forms of contract. Even the price is not certain at the outset. It is designed to be uncertain - to be left to the end and fixed by reference to what was actually done. But when forms such as these are carried into modern integrated projects - and not just large scale ones - demand arises to render more fixed a price at the outset, despite the inappropriateness of the conceptual basis of the form to such different contexts. So, FIDIC has been applied not just to uncertain price contracts but also to fixed price contracts; not just to civil engineering works in the ground of uncertain scope, but to electronics, pipelines and energy production contracts, even though those may well involve standard off-site fabrication-type components, where the pricing mechanism would have been capable of more price certainty at the outset.

Thus a central aspect of legal risk management is this: not only are the difficulties of legal risk as much a product of the institutional ways in which we handle projects, but also more to the point they are the product of the inappropriateness of choice of contract relative to specification, and relative to the project.

Degree of definition of the works

Take a critical example: how well are the works defined at tender stage? Very often it is said that one must not let a job until the works are defined. But that is frequently commercially impossible - not a realistic expectation. So the legal community and commercial community have developed innovative contractual frameworks, such as construction management, or project management (for example the New Engineering Contract). One has to make sure that the pricing mechanism that is adopted is appropriate to the degree of the definition of the works at tender stage. Of course it may be said: financiers like fixed price (or lump sum) contracts; contractors prefer cost plus; civil engineers assume a remeasurement basis. The question is: which projects are appropriate for which pricing basis?

The risk management issue is not that uncertainties are inevitable. Rather, how can one relate the realities of a particular construction project, and its recognised uncertainties to the finite commercial realities encompassing it, such as financing and completion risks? So lump sum pricing is appropriate where there is a high degree of definition of the works at tender stage. But it would be odd to apply a remeasurement contract such as FIDIC) to a project which is predominantly a process plant construction. Cost plus obviously has dangers of inefficiency, but at least it is flexible and can be made more attuned to the day to day circumstances of the project: if combined with target incentives and close monitoring it avoids the inefficiencies of contingent pricing.

Certainty of price *vs* uncertainty of design

The modern movement in apportioning responsibility is towards single-point, often with fixed price and contractor responsibility for design. But those with a financial dependence on a turnkey solution will want their own opportunities for technical monitoring. It is a major source of legal risk that requires careful contractual definition how to provide for contractor design responsibility and yet client participation in approval processes. Experience shows that many legal risks eventuate as disputes because of the inappropriateness of approaches to those two roles. Very often delay on a project arises because the approval process on the client's side is misunderstood. This is even more true when there is a mixture of performance specification and descriptive specification.

Weakness on programming obligations

Take one more example of the potential inappropriateness of traditional approaches to construction contracts: the traditional standard forms are weak on programming techniques. Of course, programmes are frequently required because of the completion risk for the financing of the project (especially where there is a revenue flow required by a particular date). Nevertheless it is frequently the case that the contractual significance of the programme is inadequately defined. The JCT family of forms, in particular, makes almost no use of programming techniques as such. The standard engineering forms, ICE and FIDIC, are not much better in this respect: a programme is provided for (in clause 14), but is not integrated by milestones, incentives, etc, into the framework of obligations of client and contractor. More recently published forms provide better models in this respect such as GC/Works 1/Edition 3 and the New Engineering Contract.

Contracts are not adversarial: disputes need not be inevitable

With better attention to the management of uncertainty, and appropriateness of contract framework, projects do not have to commence on the basis that disputes are inevitable. Projects do not have to be adversarial in nature, nor are forms of contract necessarily adversarial in nature, as is frequently suggested. What is true is that uncertainty is more likely to produce disputes. By realistic appraisal at the outset of levels of uncertainty it is often possible to remove the seeds of what would otherwise be eventual disputes.

There are ways of handling and managing risk which ought to prevent disputes arising in the first place, even during the progress of the project. The traditional contract administrator is in the position closely to monitor the project and particularly the quality systems implemented for the project. He ought to know in some detail first hand the basic facts. A Disputes Review Board, a Panel of Experts or an Adjudicator has the merit of being able to stand back more and assess what has happened. Certainly, it is necessary to find methods of keeping the dispute review mechanism (whatever form it takes) informed as to what is going on the project. "Alternative" methods of dispute resolution - not just for smaller scale projects - have considerable merits, such as mediation or conciliation, even if not inherently binding other than by agreement. Particularly while the project is continuing they enable cheap effective removal of difficulties, so that the project can progress satisfactorily and get back on track. It is undesirable that such matters should fester and become an impediment to progress.

The context, and competence, of the client

To some extent, these matters depend upon the competence or experience of the client and the competence of others in the project sharing the risk. Some suggest that among the aims underlying the traditional ICE approaches were to make the public sector client have the risk pushed away from him and have it resolved in a very accountable way which can be explained to public accounts committees. By contrast, when one looks at, for example, the process side of the industry one has, for example, large oil companies as clients who themselves are much more competent in the technology of the project. They are therefore much more capable of assessing the risks and coming to an appropriate bargain. That may explain the view that on the process side of the industry there are more effective forms of contract and more effective dispute resolution.

A further way of expressing risk laden differences in types of client is a scale between resourced and not resourced, at least in a financial sense. This may be just as significant as the relation of the product in a technical sense to the business of the procurer. So, of course, for example in the oil industry, if one takes a large purchaser of a plant, which is itself as an organisation not only extremely skilled in the previous procurement of such plants, but also in their management and general understanding of that industry; such an organisation has a more obvious "whole life cost" interest in that project. That may make the construction phase of it, relative to the whole life of it and of the organisation, further down the scale. The opportunity and revenue benefits over the whole life of the project may be much more significant. The particular project is then but part of an overall pattern of procurement within the business which itself may be international. That puts a completely different perspective on the management of risks that arise and their solution.

This latter context is in stark contrast to the typical context of the most difficult and intractable disputes. Those frequently arise where in retrospect there is a lack of knowledge as to what has really happened, or worse still, none of the participants can really afford (because of the isolated context of the project relative to the continuing life of that entity) to deal with what has now happened. It may not be fully appreciated why the problem arose but it is certainly the case that neither side can now afford to accept the cost of putting it right. By contrast, if the problems arise in a context which is reasonably well resourced and can be viewed in the context of the whole life of the project, then the chances are that a constructive path can be found out of that problem. It is also likely in the latter context that the market place will enjoy a high degree of cohesion between those buying such engineering and those providing it.

These differences can be illustrated by reference to advantages and disadvantages of Construction Management methods of procurement. Construction Management illustrates very well the difference between the experienced client who can often make very good use of construction management, and the inexperienced employer who might be much less well fitted to enter into direct contracts with many works package contractors directly.

The "open texture" of language, and approximation in legal methods

The law interfaces with risk assessment both at the stage of contract drafting, and in the carrying out of that contract. But, the contract draftsman and the ultimate tribunal are still limited by the open texture of language and the necessary approximate methods which are involved in methods for legal trials and the assessment of evidence. Legal techniques will attempt best assessments of the meaning of risk probabilities. Nevertheless, in the law, ideas of causation and likelihood retain substantial elements of impression rather than purely analytical scientific method.

The solution to such apparent shortcomings is in the identification, collection and maintenance of records. Good and cogent evidence of pertinent data will reduce levels of uncertainty in the contractual risk assessment. Much will depend on the extent and period of the tendering process and whether there is a realistic opportunity for the contractor to initiate his own investigation and substitute his own judgment, especially when such an exercise has already been undertaken on behalf of the employer.

Some conclusions for procurement strategy

The common goal is, so far as reasonably possible, to meet the client's objectives. So, the purpose of the contract framework, as a tool in risk management, is to cause - by definitions and incentives - the risks to be identified and managed by practical means most likely to achieve that goal. The contractual framework is a fundamental tool, to be designed and used creatively. Just as planning is a tool for the reduction of uncertainty and, for example, funding is a potential tool for responding to the eventuation of uncertainties.

Key words in procurement strategy are "appropriateness" and "realism". A definition of appropriate contractual provisions has to be a considered response to the particular circumstances of each project. A considered response has also to be an articulated response. Too often uncertainties are left unaddressed in qualifications to tenders, or unexpressed and conflicting assumptions carried into the project by the different participants. A typical example can be discrepancies and inconsistencies that arise between stated elements of the client's brief and expectations arising from quality standards and codes of practice. Simplistic contractual devices, such as ill considered transferring away of risk, may in fact in the end cost more money. Managing risks, recognising the principles in this overview, is likely to save money.

ANNEXE 1

The 1992 SERC project report *"Engineering construction risks"*[5] included amongst its conclusions:

- **All too often risk is either ignored or dealt with in an arbitrary way:** simply adding a 10% 'contingency' onto the estimated cost of a project is typical. This is virtually certain to be inadequate and cause expensive delay, litigation, and perhaps bankruptcy.

- The greatest uncertainty is in **earliest stages** of a project, which is also when decisions of greatest impact are made. Risk must be assessed and allowed for at this stage.

- **Flexibility in project design** and the **risk of later changes** should be considered in detail before completing proposals for sanctioning.

- Risks change during most projects. Risk management should therefore be a **continuing activity** throughout the life of a project.

- On most construction projects, the client deceives himself if he uses single figure estimates of cost and time for appraisal and funding decisions. **Ranges of estimates should be used,** including **specific contingencies** and **tolerances** for uncertainty.

---

[5]   *Op cit*, n.1

- **Delay** in completion can be the greatest cause of extra cost, and of loss of financial return and other benefits from a project. **The first estimate of cost and benefits should be based on a realistic programme** for a project. On this basis the potential effects of delays can be predicted realistically.

- Attention to contract strategy based upon systematic consideration of risk can achieve significant cost savings for a project. There is growing acceptance in the UK that traditional contractual arrangements are no longer the best basis for managing today's high -risk projects. **The proposals for funding (sanction) a project should therefore include recommendations on contract strategy.**

- **Competitive tendering** coupled with **traditional contractual arrangements** limit the realistic management of risk. The pressure is always on those bidding for contracts to keep their tender prices as low as possible, which can put both them and their clients at great financial risk if things go wrong. When some provision has been made for eventualities, it is often buried in the total bid. This hinders the effective management of risk and militates against a systematic and equitable basis of payments.

- The overriding conclusion drawn from the research is that clients and all parties involved in construction projects and contracts benefit greatly from **reduction in uncertainty prior to their financial commitment**. Money spent early buys more than money spent late. Willingness to invest in anticipating risk is a test of a client's wish for a successful project.

2. The Control of Risk

Patrick Godfrey

Synopsis

This paper introduces a simple approach to identifying and assessing risk. It is demonstrated how this approach can be extended to cost benefit assessment and performance monitoring of complex construction projects. The paper aims to make explicit the intuitive understanding of risk that is the basis for traditional risk management. Where the availability of data and cost benefit allow, this should encourage the use of more advanced methods of assessment.

Introduction

The forthcoming CIRIA publication *"A Client's Guide to Risk Management in the Construction Industry"* is based on research into how clients perceive risks in construction projects and how they approach its management. Ten "typical" clients were interviewed. The survey indicated that only a small proportion of building and construction industry clients are aware of, or have experience in the use of, systematic risk assessment methods to help them control risks to their business from their construction projects. There appears to be a willingness amongst the clients interviewed to consider the use of systematic methods, provided the benefit to be gained from their use and how to employ them themselves as part of the normal course of management is underlined.

The outcome recognised the existence of a "cultural gap" in terms of clients' approaches to control of risk with the technology available, and a strong desire to understand and adopt methods that allow them to better control their risks. The research concluded that:

- the guide must clarify as early as possible what risk management is;
- the guide must identify as early as possible the benefits of risk management; and
- the guide must lead the reader/client into making a "first pass" risk assessment of his own, so that he understands the issues involved and can, where necessary, make reasoned decisions about appointing appropriate professional advisors.

This paper seeks to address these issues

What is Risk Management?

What is Risk? Simply put, risk is the chance of an adverse event. More technically, *risk* is the combination of the probability or frequency of occurrence of a defined hazard and the magnitude of the consequence of that occurrence. Thus, risk is a measure of the likelihood of a specific unwanted event and its unwanted consequences or loss.

$$\text{Likelihood} \times \text{Consequence} = \text{Risk}$$
$$\text{or}$$
$$\text{Probability} \times \text{Cost} = \text{Risk}$$

Risks faced by clients are wider than those normally associated with the construction process. It follows that:

- Risk to a client's construction project is not the same as risk arising from that project to the client's business;
- Risk management focused solely on the project is of limited benefit to the client; and
- Action to control risk arising from the construction projects which affect the client's business, will not necessarily be the same as those employed to control risk in the construction project itself.

The construction industry tends to lose sight of the fact that it depends on its clients' prosperity for its own, and that the construction process itself poses risks as well as benefits to that prosperity. Similarly, clients lose sight that transfer of all the risk in the construction process to the construction industry means that they will not have control over the risk management facility. Projects may be managed without regard to clients' business requirements, which can change during the period of years necessary to complete most projects.

The quantification of Risk

The consequence of an adverse event, sometimes called damage, is often expressed in monetary terms. This may not always be easy or appropriate, especially when the risk concerns fatalities and the delay and cost implication. Other performance measures may be more convenient. The likelihood, or probability, of an event, is usually expressed in terms of the number of events expected to occur in a year. Thus, if the event is expected once every 10 years its probability is 0.1 per year. Time is, however, not the only consideration; for example, the likelihood of defects in production of steel bars might be expressed in terms of kilometre of bar produced. Risk, as the combination (multiple) of consequence and likelihood, will usually be expressed in £ per year.

Events that, though unlikely, are potentially catastrophic may require particular attention. The risk may appear insignificant because it is unlikely, yet it could happen. If the consequences are unacceptable then the risk must be avoided in spite of its low probability. Such events are typically those that are insured against, for example, fire.

The management of Risk

Some large companies that can afford to absorb losses actively pursue a high risk/high return strategy in their area of experience. Such a strategy can actively influence the company's competitive advantage. Smaller companies tend to spread the risk, in order to reduce its overall effect, at a lower rate of return. They cannot afford to pursue a high risk strategy because of the effect of failure. It follows that attention is concentrated on identifying what can be done after something has gone wrong, and establishing a procedure to avoid it in future. The cost benefit of such risk management action is seldom calculated. The usual approach to management of risk is based on intuition and past experience. It involves a high level of judgement.

Intuitive risk management procedures tend to be resource-efficient, but not always effective. Change is a source of risk that requires particular care in its management, because, by definition, simple extrapolation of past experience could well be misleading. All construction projects imply, or are a result of, change. However, the accelerating rates of construction, added complexity, wider variety of contractual relationships and forms of contract, increasing technical content and fragmentation of many clients through privatisation, have tended to threaten the reliability of intuitive methods. Where organisations are created for particular projects they will have had limited opportunity to build up their own risk culture, therefore clients tend to develop one-off risk management strategies.

Systematic management of risk

Where culture and experience exist in a stable low risk environment, intuitive risk management has performed reasonably well, provided it is not distorted by uninformed dogma or other mechanisms such as hidden agenda. However, where change, fear of the unknown and high hazard situations are the source of risk it is particularly beneficial if the process of intuitive learning is better informed by analysis or assessment. High hazard and high technology industries such as nuclear, oil and gas, food processing, water treatment, railways, and defence have been using systematic techniques for some time. It is relevant to note that these high hazard industries are not high risk industries. The hazard to life from railways or aircraft is higher than that on roads but the safety risk is much higher on roads. Where systematic methods have been used additional benefits have been identified from practice. They help to:

- Clarify objectives;
- Better understand uncertainty ;
- Respond better to the unexpected;
- Promote effective communication;
- Build the team;
- Improve decision-making;
- Create, reinforce and restore confidence;
- Protect the balance sheet by transferring or avoiding unaffordable risks;
- Eliminate unnecessary risk control measures;
- Concentrate resources on what matters;
- Avoid overruns;
- Release and save contingency;
- Demonstrate that risk has been managed.

With such a wide range of possible benefits there is a danger that systematic risk management techniques will be introduced too fast. As a consequence, the benefit of existing methods may be impaired before the new culture has reached maturity. The process of implementation and the outcome is counter productive.

Some of the most cost effective systematic risk management tools are designed to promote the learning process that will make them specific to the use required. It follows that an evolutionary approach to their introduction will be more effective than dictatorial, revolutionary or evangelical approaches. The methods start by mobilising judgements to identify and evaluate the cost benefit of risk reduction actions. They concentrate on practical actions and decision and help to minimise the influence of intuition and approximation. In the construction industry, where a high proportion of risk arises from uncertainty, in part due to lack of information and the prototype nature of construction, the practical bias is essential. It is, therefore, particularly important that the methods are used to inform judgement and not act as a substitute for it.

Cost and transfer of risk

Most risk issues arise from the need to control and monitor progress: "What can go wrong". Risk and responsibility are inevitably interrelated. Many clients seek to transfer the cost of risk to others, but resist the transfer of its control. Whilst this is a logical consequence of the need to protect business from risk, it does not necessarily represent the most cost effective solution. Provided a buyer for the product can be found, the cost of risk is saleable and thus transferrable: the insurance and gambling industries are built on this principle. The sale of risk is generally at an increased cost and, if the transfer of risk leads to a lower quality of management, in the long run this means increased costs for everybody. There is an undoubted economic benefit in concentrating attention on the quality of risk management, rather than pursuing a general argument as to who is willing and able to pay for it.

> *" Once a client is satisfied about real need and feasibility within overall budgetary constraints, the instinctive reaction is to retain a consultant to design the project - the "ring up an architect / engineer" syndrome. That takes a crucial step too quickly, and closes off potential procurement options. The next step should be the use of internal risk management to devise a contracting strategy. The client should decide how much risk to accept. No construction project is risk free. Risk can be managed, minimised, shared, transferred, or accepted. It cannot be ignored. The client who wishes to accept little or no risk should take different routes for procuring advise from the client who places importance on detailed hands-on control. "* [1]

The logic of this approach is undeniable. It will, however, be more difficult to implement effectively than it appears. In the same way as design and the development of contracting strategy is in practice, risk assessment should be an iterative process, except for very simple projects. It has also to be recognised in implementing the recommendation that a large proportion of clients do not at present have the means or experience available to them to undertake systematic risk assessment. It is unsound to argue that the systematic methods can rapidly be transferred from one industry to another, or, as importantly, from one profession to another. The culture gap takes time to close. If systematic risk management continues to be introduced, step by step, as skills spread and experience builds, then it should help to avoid some of the very costly mistakes that have been made in the name of contracting strategy in the UK in the past decade.

---

[1] *Constructing The Team, Final Report of the Government / Industry Review of Procurement and Contractual Arrangements in the Construction Industry*, Sir Michael Latham, HMSO, January 1994, § 3.7.

The Team

The client

The client is the key to success. He must recognise his own importance. Regardless of his construction experience, he will have detailed knowledge about his own area of work; he will know what he wants to achieve and why. The input from the client as commissioner of the construction project is vital. Therefore, when it comes to planning his scheme, he must define, in as much detail as possible:

- his objectives; and
- his constraints.

The client must have the confidence and trust of his financiers and/or equity stakeholders, and experience sufficient to allow procedures to develop without losing sight of his objectives.

Facilitator

Experience has shown that during the course of a risk assessment somebody should be appointed to the role of facilitator in the risk management team. This person may be an existing member of the project team, or somebody may be expressly appointed to take up this role. The facilitator must be proficient and experienced enough to take on this challenging and key role. As far as possible, he should have:

- ability in all relevant risk management techniques, to help command the confidence of all the parties involved;
- a good understanding of technical issues relevant to the project;
- a good understanding of commercial issues relevant to the project;
- good management skills; and

- good communication skills.

The facilitator will head the risk management team and ensure that the client and his professional team get the full benefits of systematic risk management. It goes without saying that in appointing an external facilitator, in addition to the above capabilities, a high degree of independence and objectivity is required. Such an appointee will, therefore, be able to offer impartial advice or guidance, which may be unpopular with one or more of the parties involved in the project; experience in managing risk issues has shown that fear of blame can inhibit effective future management.[2] The facilitator is in a position to encourage an atmosphere in which blame-free reporting can flourish.

Criteria for selection of the risk management team

The facilitator should offer guidance when selecting the other members of the risk management team. In accepting this guidance, the client should not lose sight of the requirements that the team should:

- aim to be objective;
- have a good understanding of the project's objectives and constraints;
- include a representative from each of the key disciplines involved;
- be as small as possible, to assist decision-making;
- be flexible; and
- be inquisitive or lateral-thinking.

---

[2] For example, air traffic control authorities have introduced systems of blame-free reporting so that pilots can report their 'near misses', unencumbered by fear of negative employer reactions.

The selection of the risk management team should reflect their ability to identify risk and develop strategies to manage it. It is the quality of the decisions taken that provides the ability to manage risk, not the size of the balance sheet available to accept it, nor the man-hours taken on the risk assessment

The Tools

Personal tools

Every risk manager will have his own perception at to the importance of particular risks. It follows that no set procedure can be appropriate to all circumstances. Indeed, since risk management is concerned with change, perception of risk has also to adapt to the needs of the time. The risk manager is like a craftsman who has various tools that with use are honed and polished to his need.

Tools for the job

Some of the tools are like jigs that need to be shaped for the particular job. It is intended that the starter pack of tools described below will be adapted to the needs of the user so that what on first use should produce a workman-like product will for some become a craftsman's stock in trade:

- Procedure can be an excuse for avoiding the use of judgement yet judgement is vital to risk management;
- Change is at the root of many risk issues. Rigid procedure can be a reflection of past requirements which will create a sense of false security in the face of the need to respond to change;
- Most organisations have plenty of procedure but few mechanisms to ensure that the procedure is appropriate to the special features of the particular issue.

Learning tools

Risk management is a learning process. What has been intuitive becomes, through implementation of the systematic approach, to an extent, explicit. This encourages feed back into the tool box in various ways:

- The 'What can go wrong' list is a concise means of helping to avoid repetition of mistakes;
- The 'Prompt' list has been developed because it has been found that these words trigger the useful risk identification debate.
- All involved in the risk assessment process should be encouraged to send a copy of their concerns whenever they arise to the risk register file for consideration at the next review; the top copy should be sent to line management at least.

Risk Identification Tools

a) What can go wrong analysis

The first step of the analysis is to decide the objective. It may be health and safety risk, risk to budget, programme or quality, or risk to any other objective or aspect of the construction project. It is sometimes useful to start the identification process from a consideration of known hazards, for example, fire, but it is generally simpler and more practical to address the risks as they stand at the time. It follows that risk identification may be regarded as a "what can go wrong analysis". By addressing this question, those topics that have the potential to harm the objective are addressed.

There are many approaches, but a systematic approach which examines the sequence of activities necessary to complete the work, may be appropriate. For example, there may already be an activity plan setting out in a step by step manner the activities which are necessary to complete the project. For each activity it is, therefore, necessary to consider what can go wrong, and make up a list that will eventually form the start of the risk register: the issues should be expressed in realistic rather than general terms.

b) Free and Structured Brainstorming

Free brainstorming simply involves recording the things that can go wrong. The process can be carried out by an individual but a better result is generally achieved by a group. Free brainstorming can be a useful warm up exercise for structured brainstorming, because it promotes consideration of the issues which are foremost in the minds of the team members, thereby allowing time to be devoted to the more subtle or unusual issues. Issues that have been a cause for concern for the team will generally have already been identified intuitively, and would probably have been brought under control. The systematic process of risk management aims to improve on this.

The use of activity lists and bar chart programmes has already been introduced as a means of stimulating the identification of risks. They are useful prompts. Other forms of 'prompt list' do however exist.[3] In each case items are selected from the list to stimulate the identification of risks to the objective.

c) Prompt lists

Prompt lists are intended to stimulate specific risk identification. They are not intended as check lists, which tend to be of limited use as risk identification aids. The prompt poses a reasonably open question. The check lists tend to be more prescriptive.

---

[3] *Total project management of construction safety, health and environment,* European Construction Institute, Thomas Telford, London, 1992.

d) Structured Interviews

It is likely that listing risks will have identified some areas of concern that require special knowledge to address. It is often sufficient to arrange to interview a suitably qualified professional. It may be more effective to structure the interview by:

- Preparing a set of open questions;
- Send the questions in advance to the interviewee;
- Agree confidentiality;
- Agree whether it is acceptable to use a tape recorder;
- Record the outcome in terms of the understanding you have gained and if necessary confirm it by sending the interviewee a copy inviting comment;
- Add or modify items on the risk register.

e) Hindsight reviews and case examples

Towards the end of a phase of work or project, it is often beneficial to the team members to meet and discuss the execution of the project. This process is generally of great benefit to those who participate in it. However, constraints of blame and claims may well inhibit wider dissemination of the experience. Nevertheless, the outcome can be used to update the prompt lists and add, or comment on, the risk register, This can then be converted back to a non-specific what can go wrong list: anonymity avoids the problem.

f) Completion of the identification process

The initial outcome of the risk identification process may appear to be somewhat chaotic and arbitrary. Therefore it is essential that the risks are described as concisely and specifically as possible in a logical sequence. The next section provides examples of risk registers and questionnaires intended to help achieve this.

Risk Registers and Risk Assessment

The risk register is a means of recording and helping to control the risk management process. The basic steps in the risk management process may include:

- Identify and record risks (essential);
- Assess or quantify the risks;
- Identify ownership (and change of ownership) of risk;
- Eliminate from further consideration but keep a record of those risks that are too small to require further attention (only possible if they have been assessed or quantified);
- Identify risk reduction actions (mitigation) where possible (essential);
- Assess or quantify or describe risk after mitigation (residual risk);
- Prepare a costed action plan to implement mitigation;
- Identify when further information may allow better assessment;
- Identify when mitigation action can be implemented.

The circumstances will generally dictate which steps are appropriate. In its simplest form the register will:

- Describe the existing risk; and
- Record possible risk reduction (mitigation) actions.

In addition, depending on the circumstances, the following amplifications can be made:

- Subdivision of risk into more detail;
- Record a measure of importance/severity;
- The ownership of the risks;

- Assessment of importance/severity of the risks, broken down into their components of likelihood (probability) and consequence;
- The capability / responsibility (ownership) for action;
- The timing of action (When);
- An assessment of the residual risk after action has been taken;
- The change in the importance/severity as a result of action;
- The estimated cost of action;
- An assessment of cost-benefit.

The description of risks and actions should be both specific to the project and concise; generalities may conceal an important issue. It is not normally practicable to compile all of this information as a single report.

Confusion may arise because, in practice, risk reduction or mitigation actions will be built into normal practice and management methods. It can be a waste of time back-tracking to the hazard itself; care should be taken to establish the state of the project at the time of the assessment. This is normally achieved by keeping a record of the meeting dates and the documents reviewed. A danger arises if a subsequent management decision eliminates a significant risk mitigation action in ignorance of its consequence. The risk register must not be used as a substitute for engineering competence in decision taking; it is impractical to identify all risks.

a) Risk Evaluation

Risk evaluation can be either qualitative or quantitative. Qualitative descriptions can be used to make a quick assessment, or may be of specific use in identifying attitudes to risk. Whether a qualitative or quantitative evaluation is required, it is helpful to establish a set of criteria that provide yardsticks against which estimates can be judged by:

- Scale points to provide importance ranking;
- Nominated values to provide risk cost estimates;
- Qualitative descriptions to establish acceptability conditions.

In each case the components of risk, likelihood (probability) and consequence (cost), can be evaluated separately. The method only expects the user to identify values within multiples of 10. No great accuracy is expected. Even though rough estimates are made these should be able to assist in the decision making process. Values should be chosen to cover the range likely to be of interest. If the scale points or the nominated values are multiplied they provide a measure of risk. High likelihood is associated with low cost; and high cost is associated with low likelihood.

b) Reflecting Bias in Tolerance of Risk

Risk evaluation may need to reflect the effect of the risk on the business. It may be that frequent marginal risks are more important than the occasional severe one. This is sometimes the case for a service industry. Passengers may be more tolerant of a 2 hour delay every six months than of a 10 minute delay on average once a week. The lost time is about the same for a daily traveller to work. Of course the cause of the delay may also influence the importance attached to it. A 2 hour delay due to bomb scare has less impact on the business than the same delay due to the wrong kind of snow!

Alternatively, it may be that the infrequent catastrophic risks are of more concern than the more frequent marginal risk. This is the type of bias that needs to be considered when evaluating safety issues. Two approaches can be adopted:

- Weighting the values of likelihood or consequence; and
- Establishing appropriate acceptability (or "tolerability") criteria that reflect the treatment of the risk.

c) ALARP

With the introduction of regulatory requirements for Safety Cases, a method known as ALARP (As Low As Reasonably Practicable) has been developed to quantify "tolerability" criteria for the risks which impinge on individuals and society.[4] Risks above a certain level are intolerable; risks below a very low level are deemed negligible. Between the two levels risk is tolerable, provided that it can be demonstrated that the cost in time, trouble and expense of reducing the risk further would be "disproportionate" to any improvement achieved. With minor adjustment, the ALARP method could be utilised as part of a safety management system.

d) Observational approach

Where the risk arises from uncertainty, considerations as to how and when it will materialise are important. Provided effective monitoring is put in place, it can be practical to take an observational approach to risk management: risk management action is adjusted to respond to feedback from monitoring. The approach can provide large savings when used to control risk that develop over a substantial period of time. Many human behaviour, durability and ground movement issues take this form. It is important that the risk management plan is prepared, and that actions can and will be taken if an adverse outcome

---

[4]   *The Tolerability of Risk from Nuclear Power Stations*, Health and Safety Executive, HMSO, 1992; *Coping with Technolological Risk: A 21st Century Problem*, Rimington, J.B., Royal Academy of Engineering. 1993

materialises. Careful planning is required. There can be considerable benefit in some pre-investment.

e) When to stop the assessment

If no risk assessment has been undertaken, assembling a register will yield some large cost benefits. This is particularly so when it is undertaken at an early stage in the project. Inevitably, the process will to be subject to a law of diminishing returns as the clearly identifiable risk management opportunities are implemented. It will become apparent that the register is either recording hypothetical risks or that its benefits are marginal. However, there are occasions when further work can be of particular benefit:

- where management is not sufficiently convinced to approve the necessary action;
- where the team undertaking the assessment is lacking in an element of experience or knowledge;
- where the risk arises from the combination of events or interaction of disciplines;
- where the risk arises for change, novelty, or serious lack of data or information; or
- where the risk arises from or depends on human attitudes.

In each case the need is to obtain a better understanding or a clearer presentation of the issues, either as part of the normal design and construction process or as part of a risk assessment brief. Options for further assessment include:

- refining the assessment by obtaining quantitative data on the risks;
- preparing decision, fault or event trees or interaction diagrams to better understand and sometimes quantify the risks;

- investigate, simulate, analyse, model, test or mock up the risk issue;
- monitor risks and review cost effectiveness, where risks are costed;
- hindsight reviews to improve management.

Circumstances will determine which approach will be effective. It is important to appreciate that:

- not all risks can be identified;
- risk evaluation is approximate and dependent on the skill of the participants;
- construction can never be risk free even though some risks can be eliminated;
- residual risks can materialise however unlikely they may be;
- risk assessment, like design, is an iterative process.

This said it can be that prolonged assessment is simply an excuse for not taking a decision. In any case it can have a negative effect by holding up the progress of the work. This has been called "Paralysis by analysis".

f) When to update the Risk Register

The nature and source of risks change during the course of a project. Detail is established and change occurs. It is therefore normal for the risk register to be reviewed at six month intervals or prior to any major decision points, for example, land acquisition, issue and award of tenders or when major changes occur or are expected.

g) Assessing when risk control measure do more harm than good

The methods described have concentrated on actions to identify and reduce risk if necessary. There will generally be a set of risk management activities in place. Some may however not be cost effective. The tools described above can be used to assess their effectiveness. In so doing it needs to be appreciated that some risk reducing measures can do more harm than good. For example, the provision of a risk control measure may create a false sense of security which increases the risk out of proportion with the risk reduction measure. An overly costly risk management measure may be preventing more cost effective measures being provided elsewhere. The risk may be moved to where it is increased by less efficient management.

The principle is not restricted to purely commercial risks but is an ethical basis for the application of ALARP to health and safety risks.

> *"We can take it for granted that too much can be paid to avert harms; my thesis is that to pay too much to avert any harm is likely to increase harm in the long run."*[5]

It is not uncommon for the "more harm than good" principle to depend upon the frame of reference. What may be a risk reduction measure for a project may be causing increased risk to the client's business. Perhaps this is part of the reason why the clients interviewed clearly identified "promoting teamwork" and "rapid resolution of disputes" as key factors in their risk management strategies.

Conclusions

Based on practical experience, the systematic approach to risk control outlined in this paper has many benefits and its development and implementation should be encouraged. Its introduction should be used to encourage the elimination of the

---

[5]   *Ibid.*

"cultural gap", rather than be forced at a pace or in a manner that widens it. It is important that those who are involved feel a sense of ownership, and keep in mind that:

- Risk assessment can only inform judgement not replace it;
- It is not possible, and is unsound to assume, that all risks have been identified;
- Risk assessment, like design, is an iterative process which should be progressed throughout the project and into operations; and that
- The observational approach to risk management can have substantial benefit and may be eliminated by premature selection of contracting strategy.

By and large, systematic risk control methods have been developed and used by high hazard industries including offshore construction. A culture gap exists between the use of these methods and the intuitive methods traditional in the general building and construction industry. In spite of this, Government regulations and policies exists, and more are emerging, that require the use of systematic methods.

The effectiveness of the use of systematic risk management methods will depend on closing the cultural gap between intuitive and explicit methods for all involved. The culture gap is greatest for those not directly concerned in practical design and construction, where probability based decisions are a matter of common practice.

3. Contract Documents and the Division of Risk

John Uff

Synopsis

This paper considers the division of individual risks in construction contracts, and the extent to which they are allocated effectively by the contract terms. The likely impact on the parties of risks arising during construction is examined. Suggestions are made as to how a contractual framework may be employed to allocate and manage risks, with particular reference to unforeseen conditions.

Introduction

Risk theory predicts the effect of risks being apportioned in particular ways between contracting parties. Management science seeks to control and minimise the impact of individual risks. In both cases, the application to real projects must utilise and be channelled through the terms of the contract. These will determine the effect which actual events have on the rights of the parties. As will be seen, the complexity of contract terms is such that a risk is rarely placed fully or simply on one party.

Risk in the present context refers to circumstances outside the direct control of the parties which have a potential impact on the project. They may conveniently be divided into the following short list quoted in the Latham Report:[1]

(1)　Fundamental risks: War damage, nuclear pollution, supersonic bangs;

(2)　Pure risks: fire damage, storm;

(3)　Particular risks: collapse, subsidence, vibration, removal of support;

(4)　Speculative risks: ground conditions, inflation, weather, shortages and taxes.[2]

The list is not definitive but indicative. One thing which it does indicate is the tendency to confuse cause and effect or perhaps risk and result. This is no doubt due to the way insurance policies are drawn, for example, treating "subsidence" as a risk when it would be more accurately described as a type of damage. When dealing with risks, contracts usually treat all risk as adverse, although some may be beneficial. For example, ground conditions or weather may be better than anticipated, inflation may fall, and perhaps even taxes.

In most instances, construction contracts do not deal with apportionment of risk as such, but with broad commercial obligations which are subject to exceptions. It is these exceptions which are usually based on specified risks. For example, the contractor's obligation to perform in due time or pay damages in lieu, is subject to exceptions including (usually) exceptionally adverse weather. Therefore, in addition to considering the division of individual risks, it is proposed to look secondly at the commercial obligations under the contract in order to see to what extent they are at the risk of one party, risk

---

[1]　*Constructing The Team, Final Report of the Government / Industry Review of Procurement and Contractual Arrangements in the Construction Industry*, Sir Michael Latham, HMSO, January 1994.

[2]　*Ibid* § 3.9, extracted from Clamp, H., *The Shorter Forms of Building Contract*, 3rd Edition, Blackwell Scientific Publications, Oxford, 1993.

in this sense meaning overall responsibility or, in insurance language, "all risks". This is indeed the way in which the term is used when referring to the "risk" of collapse or subsidence of a building. If consistent terminology were to be used, one might refer to the risk of occurrence of the individual circumstances leading to collapse or subsidence, but these usually figure as exceptions, as already noted.

There is then a third aspect which needs to be considered under the umbrella of contract documents, namely the impact of risk. In most cases risk is to be equated with uncertainty which is itself susceptible of analysis in terms of probability of occurrence. In some cases, particularly uncertain ground conditions, the probability is the product of, or at least influenced by, the actions of the parties. These may be regulated or affected by the contract terms and it is therefore legitimate to inquire how the contract can or may operate on the impact of risk.

Dividing Risk

It is a relatively straightforward matter of drafting to place the direct or immediate consequences of a risk on one party. This may be achieved by an express term, but even that may be unnecessary if the law would otherwise place such liability squarely on one party. However, to transfer a risk fully, it will be necessary to consider all the possible consequences and to make appropriate provision for them, using devices such as indemnities. For example, if it is desired to place the risk of adverse ground conditions on a contractor, there may be an express term to this effect. The law will ordinarily impose the same result,[3] but where ground investigation data is provided an express clause may well be necessary to avoid the possibility of

---

[3]   *Bottoms v. Mayor of York* (1892) HBC 4th ed. Vol.II p208

part of the risk being effectively transferred back to the employer. If adverse ground conditions did occur, the contractor might then become entitled to an extension of time, and it would also be necessary to ensure that such a right was removed if the particular risk was indeed to be fully transferred. Even then the employer would ordinarily receive only liquidated damages in lieu of his actual loss, and even that right might be defeated by some technical objection. Only an indemnity would insure the full transfer of risk.

In practice, the way in which such risks are spread under practical construction contracts is complex and the contractual terms often do not readily demonstrate which party is to be regarded as bearing a risk. The difficulty in achieving a clear transfer or placement of risk is due to the complexity of the conditions including the exceptions and limitations invariably found within contracts. This often means that, in addition to the problems of achieving a transfer of the risk, the transfer operates only in particular circumstances which may have little to do with the character of the risk. Partial acceptance of risk is well-known in insurance contracts where the insured person may accept an excess or deductible sum, thereby carrying the first, lower level of financial risk, while the insurer carries the higher risk. This is often reflected in construction contracts, for example, where the contractor bears the (lower) risk of ordinary adverse weather and the employer bears the time element of the (higher) risk of exceptionally adverse weather. The relative degree of risk here refers to its impact or effect, not its likelihood.

The problem involved in some elements of risk transfer can be expressed in terms of the reaction of the parties and their incentive to minimise the impact of risk. The New Engineering Contract, which recently received handsome support in the Latham Report claims that the valuation of "compensation events" should be such that the contractor is indifferent as to the outcome so that the choice can be made solely in the interests of the employer. It is to be doubted that this can be claimed of the ICE conditions and it is proposed to examine both forms of contract to demonstrate the extent to which such indifference to such events is a reality.

The majority of the "compensation events" under NEC[4] are not risks in the ordinary sense in that they are within the employer's control, for example, changes to the work or later instructions. The following risks are included within the list:

(12) Physical conditions within the site other than weather conditions, which at the contract date an experienced contractor would have judged to have such a small chance of occurring that it would have been unreasonable for him to have allowed for them;

(13) Weather which is expected to occur less than once in 10 years;

(14) Employers' risks, (Clause 80.2), viz.
 - Third party claims for injury to person or property;
 - Damage to the works;
 - War or radioactive contamination;
 - Other risks stated in the Contract Data (these are subject to exceptions).

The above is a rough summary; the actual risks and exceptions are defined in language similar to that found in other contracts. The most interesting of these risks, in terms of impact as well as the complexities involved in its transfer, is that of adverse physical (ground) conditions. Particular attention will be paid to this aspect of risk.

---

[4]   Clause 60

Risk of Adverse Ground Conditions

Unforeseen adverse ground conditions are one of the most notorious causes of disputes under engineering contracts. Their incidence may have far-reaching effects on the course of the works and their resolution can often have a most serious effect on the economic balance under the contract. Given their importance, one would expect to find a rational scheme of transfer and placement of this risk, whether or not achieving the goal of "indifference". Unfortunately, experience suggests that the effect of these clauses is anything but rational.

It will be noted that the risk defined in NEC at (12) above is remarkably similar to that defined in clause 12 of the ICE Conditions of Contract, the Sixth Edition of which uses the following description:

> *"Physical conditions (other than weather conditions or conditions due to weather conditions) or artificial obstructions which conditions or obstructions could not ... reasonably have been foreseen by an experienced contractor."*

It is not proposed to comment on the relative merits of the drafting. It is clear that the NEC version is simply a reworking in language which appears somewhat tighter, but otherwise achieves substantially the same effect, namely that the contractor is to be compensated (and therefore is not to assume the risk) for an occurrence which objectively would not be regarded as material.

The wording of the clause, in either case, gives little guidance as to when the provision applies and when it does not. It poses a test which is related in part to what has immediately occurred, but is also dependent on conditions which existed at the date of the tender, and which are likely to be obscure and highly susceptible of dispute at the date of the occurrence. The important point is that the contract terms make no attempt to define the occurrence of risk in terms which can be applied directly or readily. Both forms of contract assume that the Engineer (or in the case of NEC, the Project Manager) will give a timely decision on whether the claimed conditions have been encountered, or in the words of the NEC, whether the compensation event has happened. Experience, however, suggests otherwise.

Unforeseen conditions typically produce large claims which remain in dispute after completion. The principal reasons for this are, first, the lack of any clear criteria for determining whether the relevant events are established, and secondly, the qualified right to be reimbursed in respect of all additional cost if the event is established. The effect of these clauses in practice is, therefore, not to transfer risk but to provide a vehicle for making a claim for additional payment in the event that the relevant facts can be established subsequently. Indeed, it may be said that the clauses embody two risks namely:

(i)   that circumstances will arise which allow the contractor to bring a claim; and
(ii) that an arbitrator might subsequently find the claim proved.

Far from being indifferent as to the occurrence of this particular compensation event, the contractor is strongly motivated, in the event of difficult ground conditions being encountered, to find conditions falling within the words of the clause. Knowing that the ability to recover additional payment is dependent upon evidence and records, the contractor must proceed accordingly and the Engineer (or Project Manager) will be under a corresponding duty to the client to discover evidence to the contrary conclusion. There is little incentive for the parties to agree even upon the nature of the conditions which are being encountered, because neither party at that stage knows with what data they are going to be compared in order to decide upon the validity of the claim. The NEC contains, in clause 60(2), a check-list of matters assumed to have been taken into account, but this unhelpfully includes "other information which an experienced contractor could reasonably be expected to have or to obtain". The result is inevitably a deferred claim which can hardly contribute to proper management of the impeding events which may or may not be finally judged to be at the risk of the employer.

Unforeseen Conditions under other Contracts

Clause 12 (or clause 60.1(12) of NEC) is so well known in the construction industry[5] that contractors carrying out any form of groundwork tend to assume its existence. Under the JCT form, however, a totally different regime applies. This is of some topicality in view of the suggestion in the Latham Report that an industry-wide form be generated in place of the ICE and the JCT forms. This area of fundamental difference will need to be addressed and resolved.

Under the JCT form, ground conditions are solely at the contractor's risk but subject to the possible effect of the Standard Method of Measurement, in accordance with which the bills are deemed to be drawn[6]. Two provisions in the Standard Method are of particular relevance, namely the requirement to measure rock[7] and running sand[8], *i.e.* sand subject to a high external water table which becomes unstable when excavated. In neither case is the term precisely defined, but the comparatively low level of incidence of such claims means that the issue is not a critical one for the industry. What does cause surprise is where

---

[5]   clause 12 of the international FIDIC conditions is to a similar effect
[6]   JCT 80 Clause 2.2.2.1
[7]   Clause D6(e)
[8]   Clause D6(d)

a groundworks contractor encounters genuinely unforeseen and unforeseeable adverse conditions not falling within either of these classes, to find that the resulting additional cost and delay is entirely to his account, whereas the ICE Conditions or NEC, the unforeseen conditions would lead to full compensation with profit.

In one such case, a piling sub-contractor, having considered a site investigation report showing London clay at considerable depth, quoted for under-reamed bored piles of a standard figuration. When the under-reaming operation was carried out, the bottom of the pile shaft collapsed, wholly unexpectedly. The operation was repeated several times with similar results. The sub-contractor asked for instructions and suggested an additional number of straight-bored piles, necessitating redesign of the pile caps and ground beams. The main contract was under the JCT form, where the architect gave a permissive instruction, but holding the main contractor liable for the additional cost and delay, which was subsequently passed on to the sub-contractor. The case (which is unreported) settled with the sub-contractor contributing part of the additional cost. The site of the works was Colchester, Essex. It was subsequently discovered that collapse of under-reaming had been encountered previously and written up in the technical literature.[9] There might have been an interesting debate on whether this phenomenon was one which could have been foreseen by an experienced contractor. But under the form of contract utilised, the question was irrelevant.

Contract Obligations and Their Risk

Up to this point risk has been addressed as an individual event or circumstance external to the contract which affects one or both parties in the performance of their obligations.

---

[9]   ICE Conference on Large bored piles, 1963

The second way of looking at risk is to consider the object or obligation which is "at risk". This is the way in which most commercial contracts tend to be structured. The primary elements of any commercial transaction (supply of goods, payment, etc.) will be undertaken by one of the parties, who will necessarily assume all associated risks unless relieved by law or by a provision in the contract. Instead of concentrating on individual risks (which are rarely placed wholly on one party) it is proposed now to consider some elements of construction contracts which are broadly at the risk of one of the parties, to examine to what extent the other party acquires some of the risk. The elements considered are: the works under construction, their design, and the contractor's economic interest in the project. These are some of the key elements in most construction undertakings. Other elements can be addressed in a similar way.

The Works

The question of risk and damage to the works is dealt with under the standard forms in a variety of ways usually involving a conditional transfer to the contractor who is required to maintain insurance, usually in joint names. The relevant conditions under the ICE form of contract, 6th edition, are that the contractor is rendered absolutely liable[10] save in the case of the Excepted Risks which include (i) use or occupation by the employer (ii) fault defect error or omission in the Engineer's design (iii) riot, war, rebellion, etc. (iv) radiation, pressure waves etc. The contractor thus takes all other risks, including damage to the works by fire and storm.

In contrast to the treatment of unforeseen conditions, the incidence of risk to the works is reasonably clear and ascertainable and is in any event unlikely to affect performance

---

[10] Clause 20(1)

of the work. The contractor may be required to rectify any damage whether or not he is liable. Since the contract requires insurance in joint names[11], the division of risk between the contractor and the employer is comparatively unimportant. What is of more importance is the Excepted Risks, which the employer alone bears.

The difficulty with these provisions, as with many similar issues which arise under the contract, is to distinguish between rival causes of damage where, for example, the loss is attributed alternatively to a design defect (an Excepted Risk) or to storm damage. What is not at all apparent from the contract clauses is the way in which such a question should be resolved. The legal test is to ascertain what was the effective or dominant cause of the damage: if this is within the exceptions then the party otherwise at risk (and the insurer) is entitled to escape liability[12]. The same result will apply where it is found that there are two causes of the damage, one the subject of an exception. While it would be possible, in principle, to draft a contract so as to make clear the precise circumstances in which the contractor would not be liable, it must be remembered that these provisions are dependent in practice on the availability of insurance and it is the insurer who will ultimately dictate the terms upon which risk can be accepted. Disputes of this sort usually take place post-completion of the project, and therefore have little or no effect on management of the works. Such a dispute may, however, have a fundamental effect on the financial viability of the project in the event that the particular damage is not covered.

---

[11]   Clause 21
[12]   *Leyland Shipping Company Limited v. Norwich Union Fire Insurance Society Limited* [1918] AC 350; *Wayne Tank & Pump Company Limited v. Employers Liability Assurance Corporation Limited* [1974] QB 57.

Design of the Works

The question which party assumes the risk of design arises under virtually every construction project, including those referred to as "Design-and-Build" contracts, since even here the client must specify his own design requirements. The term "design" has no precise meaning in the context of construction projects. All construction involves elements of design and all participants in the production of construction works are, in some degree, "designers". Design represents a spectrum of activities within the overall construction process. It embraces at least the following issues and sub-issues:

(1) Function and form:
   • fitness of works
   • appearance
(2) Detail of work:
   • obligation to produce
   • approval of details
   • compatibility with other details
(3) Choice of materials:
   • specified materials
   • performance specification
   • delegation of approval;
(4) Method of work:
   • buildability
   • design of temporary works
   • safety of work during erection
   • safety of completed works;

There is large scope for dispute and uncertainty as regards liability or risk in regard to the many choices and decisions embodied in the above list. This is likely to involve potential disagreement between the primary designer (the Architect or Engineer), the main contractor and various sub-contractors, particularly bearing in mind the likelihood that design warranties will have been given. In contrast to the risk to the works themselves, risk regarding design is likely to have an impact during the currency of the works. There are instances in which parties have reached an impasse, for example with the contractor not prepared to carry out changes to the work without being given a variation instruction, and the Architect not being prepared to give an instruction other than to remove defective work[13].

There is a need to ensure that risks in the design are generally assumed by one party who can initiate appropriate action in the interests of the project when the design is found to be deficient. The only party who could undertake such liability and risk is the contractor, who is then in a position to pass on his primary liability to designing sub-contractors or to the designing architect or engineer by way of contribution proceedings. English law has been moving in this direction for many years,[14] and the French Civil Code has gone further, rendering the contractor generally liable for the fitness of a structure.[15] Generally it is the contract terms which operate to prevent the assumption of risk which would otherwise be imposed under English law.[16]

Contractor's Economic Interest

This rather odd expression is used because it represents a major element of construction contracts which is seemingly treated as

---

[13] Consider, for example, the case of *Holland Hannen & Cubitts Northern Limited v. WHTSO* (1981) 18 BLR 80 (QBD); (1987) 35 BLR 1 (CA), the Rhyl Hospital case.
[14] See *Young and Martin Limited v. McManus Childs Limited* [1969] 1 AC 454 and *IBA v. EMI & BICC* (1980) 14 BLR 1.
[15] French Civil Code Art 1792
[16] See Rhyl Hospital case above and consider *Gloucestershire City Council v. Richardson* [1969] 1 AC 480.

being "at risk". While obviously at the primary risk of the contractor himself, it is treated, under some forms of contract at least, as being subject to exceptions where the employer assumes the risk of damage to it. This is the effect of the so-called "loss and/or expense" claims increasingly found under conventional building contracts, as well as unforeseen conditions claims.

It is not suggested that contractors are paid excessively in respect of these claims. The point which is made is that contract drafting tends increasingly to involving the employer in the economics of carrying out the works, which inevitably leads to complex disputes. By contrast, a provision for additional payment to be ascertained on a direct scale unrelated to the actual economics of the work is to be preferred as being simpler and preserving the existing economic risk under the contract, whether beneficial or detrimental. There is indeed an interesting contrast between the approach to valuing variations under the JCT Conditions and the ICE Form of Contract. Under the former, the contractor is permitted to claim loss and/or expense in addition to the prices fixed in accordance with the contract, on the ground that the contractor has suffered loss "for which he would not be reimbursed by a payment under any other provision of this contract".[17] By contrast, Clause 52 of the ICE Conditions provides for valuation of variations exclusively by reference to the existing rates so far as reasonable. The nearest equivalent to a loss and/or expense claim is the provision for re-fixing of rates[18], but it is axiomatic that the starting point is the existing rates which are to be adjusted in accordance with the new circumstances. The parties thus remain bound by the contractor's quoted prices. The New Engineering Contract adopts an approach based on "actual cost",[19] although the

---

[17]   JCT 80 Clause 26.1
[18]   Clause 52(2)
[19]   Clause 63.

contractor may be asked to submit quotations.[20] A requirement for quotations for extra work reflects the approach of US construction forms, where "Change Order" procedures are more or less rigidly enforced. Under these procedures, the supervisor may not order a change unless and until the new price and effect on programme have been agreed.

Risk and Uncertainty

The third aspect of risk mentioned above, is the element of uncertainty and probability, concepts well understood by engineers when seeking to harness the forces of nature. The whole training and ethos of engineers is in terms of predicting, regulating and avoiding the impact of such risks. Adverse weather, tides and water currents are matters ordinarily placed at the contractor's risk and the means of successfully overcoming them are an essential part of the contractor's expertise. The contractor's risk may be ameliorated by measurement provisions which seek to reimburse the contractor by reference to the work required, for example, including temporary works. There is, thus, some degree of risk sharing in terms of payment as well as extension of time and release of potential damages for exceptionally inclement weather.

Where such physical difficulties impact on the works, it is a frequent occurrence that the contractor requests instructions from the Engineer which are not forthcoming. This may relate to suggested changes to the method of working, to the temporary works or even to the permanent works. It is an unfortunate feature of the standard forms of contract, particularly the ICE Conditions, that while they empower the Engineer to give a very wide range of instructions[21], they may also render the employer

---

[20]   Clause 62.
[21]   See particularly clauses 13(1) and 51(1).

potentially liable to pay additional sums to the contractor[22], and may give rise to serious disputes. Similarly, in regard to unforeseen ground conditions, Clause 12 of the ICE Conditions contains elaborate machinery for the Engineer to direct the course of the works in order to minimise the impact of the conditions encountered. Yet the frequent response to notices under Clause 12 is a denial or non-admission with the relevant decisions usually being made after the event when it is too late to affect the course of events.

The way in which the impact of uncertain conditions can be regulated has been the subject of an analysis applying risk management techniques to ground engineering methods. In their paper "Site Investigation and Risk Analysis", Peacock and Whyte[23] demonstrated how a project management computer programme CASPAR[24] can be used to calculate the probability of cost and duration of a project based on limited information. The project considered was a building the foundations of which required a mine infill scheme, the extent of which had to be predicted from limited site investigation data. The work demonstrated how the likely range of cost and duration could be refined by additional site investigation data, thereby substantially avoiding the risk of additional cost and unprogrammed delays to the project.

Peacock and Whyte did not go on to consider what incentive the parties would have under a typical form of construction contract to carry out the works in the logical and economic fashion described. Crucial questions are who is to pay for the additional

---

[22] See clauses 52, 13(3) and *Simplex Contract Piles Limited v. St.Pancras BC* (1950) 14 BLR 80, not followed by Judge Newey QC in *Howard de Walden Estates Limited v. Costain Management Design Limited* (1991) 55 BLR 124.

[23] Proceedings Inst.Civil Eng. Civil Engng 1992, 92 May 74, Paper No. 9863

[24] Computer Aided Simulation for Project Appraisal and Review.

investigation and who is to pay for extra work and delay. Only where one or perhaps both parties have an incentive to carry out additional investigation during the course of the project is it likely that the work can be so logically organised. So long as a contractor has the possibility of being paid substantial additional sums in respect of unforeseen ground conditions, there is little incentive upon him to avoid their consequence, either in terms of additional cost or delay. In the situation postulated by Peacock and Whyte, it would require a specially drafted contract to ensure that the contractor carried the risk and therefore had the incentive to minimise the cost and delay involved. Alternatively, there would be a clear incentive on the employer to carry out additional ground investigations at pre-tender, but this would not preclude the contractor making a claim and the question of risk in regard to unexpected conditions would remain. It is indeed difficult to see how management techniques can be sufficiently utilised other than under contracts based on cost-plus payment.

A Rational Division of Risk in Ground Conditions

Having now considered individual risks and their impact, it is relevant to consider whether a rational division of risk can be devised in respect of adverse ground conditions, representing an area in which both parties should benefit from minimising the impact of such a risk. What are the principles upon which a rational system of risk placement or transfer should be based? The first, it is suggested, is that the risks must be readily identifiable and the question of who bears that risk easily and directly ascertainable. Secondly, there should be a proper reason for transferring the risk, such as the party assuming the risk being better able to control it. Thirdly, the purpose of transfer should be to alleviate unreasonable loss, not to generate additional profit.

Judged by these criteria a case may be made out for transferring the risk of unforeseeable conditions to the employer, on the footing that he decides upon the initial ground investigation which ought to minimise or avoid the risk. It may also be argued that the employer or the Engineer on his behalf, is in a better position to manage the risk by giving appropriate instructions, for example, changing the work to achieve the most economic solution in the new conditions. However, this reasoning falls to the ground unless it is possible readily and quickly to ascertain whether the risk has in fact been transferred. If this is not known, the parties are left to rearrange their affairs not knowing who is to pay for the decision being taken. The result, as already discussed, is often that the Engineer makes no decisions at all thus negating the whole concept of risk transfer.

How can the existence of a transferable risk be more readily determined? There are many answers to this. One solution is suggested in the CIRIA Report on Tunnelling[25], which advocates "reference conditions" upon which the tender is deemed to be given, so that any different conditions can readily be identified and accepted or agreed. A less satisfactory but perfectly workable alternative is for the employer to warrant the ground information. This would have the added bonus of providing the clearest incentive on the employer to obtain sufficient and accurate ground data so that the information could be warranted as representative of the site. This might give rise to disputes but it would be a considerable improvement on the present clauses.

The next clear improvement would be to avoid excess profit (or loss) by regulating the claim to be allowed. It has been suggested that one means of avoiding the draconian impact of ICE clause 12 would be to transfer the risk in part only by providing (or allowing the contractor to tender) a percentage of the additional cost to be recovered. This would provide an

---

[25] CIRIA Report R79: Tunnelling - Improved Contract Practices, 1978

incentive (depending on the percentage figure) to the contractor to avoid either encountering adverse conditions which he might suspect to exist, or incurring additional cost, the full extent of which he would not be entitled to recover. The principle of risk sharing in this way could apply equally to "unforeseen conditions" clauses based on reference conditions or on warranted site reports.

Another alternative occasionally found in excavation contracts, such as tunnelling projects, is to treat the whole question of ground conditions as one of measurement, providing items to be priced covering all conditions which might be encountered. Providing that the pricing is properly balanced in order to compensate but not reward, it might then be said that it becomes a matter of indifference whether or not such conditions are encountered. It should also be possible to devise a corresponding tariff for time allowances necessary to overcome adverse conditions so as to avoid disputes in regard to delays also.

A combination of the above measures ought to be capable of producing a clause which adequately and rationally distributes the risk in the ultimate interests of efficient and timely completion of the works as well as paying the contractor a proper price without needing to generate a dispute.

Conclusions

Contractual risk is a complex, many-sided issue which has not been adequately analysed from the viewpoint of contract drafting. While the principles of risk placement may be clear, the way in which this is brought about in real contracts is usually complex. It is often difficult to ascertain which party is carrying what risks and in what degree.

As an alternative to analysing individual risks, it is informative to consider the basic obligation created by a contract and to consider to what degree they are placed at the risk of one party or the other. As regards the safety of the works under construction, this is usually dealt with adequately in terms which mirror the insurance policy required to be taken out. As regards responsibility for design, there is no clear assumption of risk; likewise the risk of additional cost occurring to the contractor is typically the subject to many contractual provisions which necessarily lead to uncertainty and dispute.

As regards the impact of contracts on the control and management of risk, there is little evidence to suggest that current forms of contract create any proper climate of incentive to operate projects on a logical basis.

Consideration needs to be given to the impact on the project should a risk arguably occur. In this event the risk should be dealt with in such a way that the parties are motivated to minimise the incidence as well as the impact of the risk. To this end financial compensation should be such as to create a situation of indifference and certainly not a situation of excessive profit or loss. A risk event should not be regarded as equivalent to a claim situation created at the will of one of the parties. While the contractor may legitimately expect proper compensation for variations and imposed delays, risks should be dealt so as to preserve a proper incentive to minimise their incidence.

Claims in respect of unforeseen conditions under ICE Contracts typify the lack of any systematic or logical approach to the placement or control of risk. The incidence of such claims is likely to have a distorting influence on the action both of the Engineer and of the contractor. Such provisions will continue to result in uncertainty and disputes unless a radically different approach is taken. That contained in the current New

Engineering Contract has nothing new to offer. The possible adoption of the NEC as a model for the industry dictates that these problems must be addressed with a fresh mind.

4. A Funder's View of Risks in Construction Projects

John Scriven

Synopsis

This paper examines the approach of funders[1] to risks in construction projects. The differences between the economic interests of a funder and those of a property company are analysed in terms of their perceptions of risk and how their objectives are satisfied. The nature and effect of construction project structures and contract terms are analysed in this framework. Conclusions are made as to future developments.

Introduction

Funding takes many forms and could include an institutional investor (for instance a pension fund or unit trust) forward funding a development where the project company enters into the building contract and the funder acquires the development on practical completion. More typically, it is a syndicate of banks lending in a "limited recourse" financing, that is, where the recourse of the banks for the repayment of their loans is to a large extent limited to a single purpose

---

[1] Throughout this paper reference is made to "the funder", although loan facilities may be syndicated among a number of banks with, typically, one bank acting as the "agent bank" on behalf of the other members of the syndicate.

project company and the assets and contracts of the project. The common feature is that the funder has a commercial interest (to a greater or lesser extent) in the completion of the development on time, to budget and free from defects, but does not itself enter into the construction contract. However, what distinguishes most funders from the entity carrying out the development (and this would include banks but not an institutional funder which will own the property) is that the funder has no equity interest in the development but lends money on the basis of a fixed return.

Although the funder's detailed requirements can differ greatly depending upon the nature of the project, this paper looks at matters which are common to a wide variety of projects. These would include the construction of commercial developments, such as shops and offices, as well as projects for industrial or infrastructure facilities such as power stations, roads and bridges. In this paper reference is made to the entity which is the company or vehicle being funded and is the employer under the building contract as the "project company". This term is normally used in connection with industrial and infrastructure project finance transactions, although this chapter covers a wider variety of projects.

In some projects the company which has an interest in the land is not the employer under the construction arrangements for commercial, tax or other reasons. In these circumstances a funder will be concerned to ensure that none of the parties to the construction arrangements can raise the argument that the employer has suffered no loss which, if upheld, could reduce the value of any claims the employer might have under the construction contract. Although the recent Court of Appeal decision in *Darlington*

B.C.v. Wiltshier[2] gives some encouragement that the "no loss" argument would not be successful in many cases, it might in some circumstances still be available to a contractor unless the employer has an obligation to the owner of the land to ensure that the works are carried out.

To some extent the requirements of a funder will mirror those of the project company. Like the project company, the funder is concerned to minimise the construction risks of money, time and quality. Like the project company, it will also be concerned to ensure that the specification and performance requirements in the construction contract reflect accurately the commercial needs of the project. These might include contractual obligations to be fulfilled by the project company in relation to third parties. For example, in an office development the construction contract may need to take account of the provisions of a building agreement with the owner of the land, or the requirements of a tenant of the completed development. In the case of a power station, the contract will need to contain performance tests and other provisions dealing with the output and other technical requirements of the plant including, for instance, liquidated damages for failure of the plant to fulfil performance criteria.

However, despite this substantial identity of interest with the project company, the funder's risk analysis and its requirements for effective security in the assets and contracts of the project are likely to give rise to significant differences of approach between the funder and the project company. It is these differences, rather than those matters which are common ground with the project company, which are the subject of this paper.

---

2   *Darlington Borough Council v. Wiltshier Northern Limited* [1994] CILL 956

The Funder's Risk Analysis and Objectives

The Nature of the Risk

The fundamental difference between the interests of the funder in a project and the interests of the project company is that the return of the funder from the project is likely to be fixed by reference to the relevant interest rates over the life of the project. It is unusual for a funder to have an equity interest in the project company or the project although this does happen in some Build, Operate and Transfer ("BOT") projects (which I mention below). In simple terms, the funder needs only to ensure that the project company is able to repay the principal of the loan and the interest, from the income of the project and/or a sale of its assets. Unlike the shareholders in the project company, the funder will not benefit beyond this fixed return. Thus it will not, for instance, share directly in any reductions in the cost of the project achieved by the project company. Except, therefore, where the cost savings can be linked with the ability of the project to repay the loan, the funder will not necessarily be sympathetic to reductions in the cost of the project proposed by the project company, particularly where the cost savings result in increased risks for the project company.

On the other hand, where the project fails in the funder's terms (that is, where the project company fails to repay the funder's loans) - this might occur for a variety of reasons, for instance due to a collapse in the market, rather than any failure connected with the construction arrangements themselves - the funder may lose a substantial part of its investment. If the project does fail, the funder will also incur the expense and administrative difficulty of enforcing and realising its security. This may be particularly onerous if the project fails before the construction is complete. All this will occur in the context of a transaction where the funder's fixed return, and its own accounts, do not take into account the risk of loss of the funder's capital.

For these reasons, therefore, which arise from circumstances outside an analysis of the construction risks themselves, the funder is likely to be averse to risk in relation to the construction arrangements. As a result, it may even require the project company to incur what its shareholders might regard as unnecessary additional expenditure to reduce the risks undertaken by the project company.

The Project Company, Equity and Shareholder Support

The project company is central to the funder's risk analysis since in general terms it will bear all the risks which cannot be passed on to the other parties in the project under its contractual arrangements with them, or cannot be covered by insurance.

In a limited recourse project, the funder's recourse is largely limited to the project company and its assets. A risk borne by the project company is therefore, to some extent at least, a risk borne by the funder. However, the risks borne by the project company can be reduced by equity or subordinated loans which the shareholders of the developer invest in the developer and by any direct guarantees which the shareholders give to the funder. Guarantees from shareholders in relation to construction may take the form of a completion guarantee under which one or more of the shareholders agree to contribute additional equity or subordinated loans equal to the amount of any cost overruns in achieving completion. The funder may also wish this guarantee to extend to other losses, such as loss of business, resulting from late completion which are not recoverable from the contractor by way of liquidated damages for delay under the construction contract, but this is likely to be difficult to achieve. Any such guarantees effectively reduce the "limited recourse" nature of the financing.

The Funder's Attitude to Risk

In its risk analysis the funder will first seek to ensure that each risk has been clearly taken by one of the parties involved in the project. Secondly, the funder will wish the risks taken by the project company to be minimised, although this might involve costs (of insurance, for example) which the project company and its shareholders think to be uneconomic. Finally, where a risk is to be taken by another party, such as a construction contractor, the funder will need to be satisfied that the party concerned has the resources to bear the additional cost arising when the subject of the risk occurs. The way in which risk analysis affects the terms of the construction contract is discussed below.

The Funder's Attitude to Documentation

Project companies and contractors should bear in mind that a funder will assess a project as a whole and will not have hard and fast rules in relation to a particular point, either relating to the project documentation on the one hand or relating to its own loan and security documentation on the other. In many cases, the strength of the contractual documentation will be assessed in the light of an appraisal of the technical aspects of the construction arrangements. Furthermore, the construction risks will only be a part of the overall risk in the project which the funder will be analysing. The funder's view of other risks may well affect its requirements for security in relation to the construction contract, on the grounds that the greater the risk that the project might fail (for whatever reason) then the stronger should be the funder's security.

Although a funder may analyse technical aspects of the project, it will probably rely much less than the project company and its shareholders on knowledge of the management of the project, including such matters as quality assurance procedures. On many projects it may have limited time to assess the project or it may not be funded by the project company (which will pay its costs, directly or indirectly) to conduct an extensive review. Where a funder does examine these matters, it is still likely to be conscious of the fact that the day to day management of the project is still necessarily well beyond its direct control. It is unlikely to have the comfort of knowing and being able to influence directly the companies, firms and individuals involved in the execution of the project. Moreover, it will be aware that, even where it has looked at the management of the project, a change in the team assigned to a project by a contractor can make a difference to the way in which the project is carried out.

For these reasons, the funder will probably attach prime importance to contract documentation and may not therefore analyse the contractual risks in terms of theories of risk management which focus on non-contractual matters. However, as mentioned above, the funder is still likely to look at the contract documentation in the light of an appraisal of the technical aspects of the project, which may affect its view of the legal obligations required. For instance, its view on the length of an exclusive defects liability period in a power station project may be influenced by the report of its technical adviser as to the extent to which the technology is new or proven.

There is another reason why a funder tends to concentrate on contractual terms. As mentioned above, the project may fail for a variety of reasons and the likelihood of the success of a project, even in relation to construction matters, may not depend upon the quality of the contractual obligations. However, where the project fails, for whatever reason (whether due to a failure of performance of the construction obligations or otherwise) a funder enforcing its security and attempting to complete the project and realise its investment is likely to benefit from strong loan and security documentation. An example of this is the requirement of purchasers from receivers of commercial developments for collateral warranties and the sometimes high commercial value placed upon them, even though a lawyer may have some scepticism about the practical difficulties of enforcing rights under the warranties. The funder will, therefore, be looking to be fully protected by comprehensive and carefully drafted (and therefore certain) contractual provisions in the loan agreement and its security documentation, since these will largely determine its relationship with the project company and its rights in relation to the project should the project fail. The funder will tend to adopt the same approach to the construction contract and the other project contracts, even though the project company might argue that different (and from its point of view more commercial) considerations should apply.

Loan and Security Documentation

Structure of the Financing Arrangements

The structure of the financing arrangements will be tailored to the requirements of the project and may (particularly in international projects) involve a number of different layers of financing and categories of lenders lending on different terms and with different priorities in terms of access to the security provided by the project company. In international financings the finance may include some loans supported by export credit agencies, for instance, in the form of a subsidised rate of interest or a guarantee of loans made by commercial funders.

Security

A key issue will be the nature of the security given by the project company to the funder. Where the project company defaults on its repayment obligations, the funder will wish to be able to take over the project and dispose of it to a third party. This will involve having appropriate security interests in all the assets and contracts of the project company required to carry out the project. The adequacy of this security will depend to some extent upon the local law (particularly insolvency law) applicable where the assets are located and the law governing the contracts.

The Funder's Technical Adviser

The funder may engage a technical adviser to check the technical content of the construction contract and to advise generally on commercial and technical issues in relation to the construction arrangements.

Many of these advisers are relatively small firms or companies and they may seek to restrict their liability, perhaps to the level of their insurance, or even largely to exclude liability for defective work by limiting the remedies of the funder to a right only to require re-performance of the services. It may be important for the funder to get the process of appointing the technical adviser under way at a very early stage and to ensure that the key questions arising in relation to the appointment of the adviser are identified early to avoid having to accept disadvantageous terms at a time when it is too late to consider an alternative adviser. However, where the adviser is in breach of its duty, the funder's effective recourse against the adviser is still unlikely to be sufficient (either because of the terms of the appointment or because of the size of the adviser and the level of its professional indemnity insurance) to compensate the funder for all the loss which it might have suffered.

Construction Undertakings in the Loan Agreement

The loan agreement is likely to contain a large number of detailed construction undertakings on the part of the project company in relation to the way in which the project is carried out. These undertakings may include an obligation not to alter the project contracts without the consent of the funder, obligations to enforce them, and also rights for the funder's technical adviser to inspect the progress of the project, to be given a wide variety of information and to monitor progress payments and completion of the project. For the project company, it may be important whether a waiver of these undertakings can be given by an agent funder on behalf of the syndicate funders or whether a majority consent of the syndicate funders is required, since the process of obtaining majority consent can be time consuming. Breach of the undertakings may give rise to the right of the funder, usually after appropriate cure periods, to accelerate repayment, call a default and enforce its security.

Completion under the Loan Agreement

Where, under the loan agreement, shareholder guarantees or support depend upon a definition of completion of the project having been achieved, then the funder's technical adviser is likely to have a role in certifying that this completion has occurred. This definition of completion may be different from the definition of completion under the construction contract and may even (particularly in industrial projects) include tests relating to the operational and commercial viability of the facility.

Requirements of Export Credit Agencies

In international projects in the developing world, insurance cover, guarantees and loan facilities may be available from export credit agencies depending upon the sourcing of procurement under the construction arrangements. Complying with the requirements of these bodies can be difficult. The bank and the project company will wish to identify at an early stage which agencies will be involved and their detailed requirements so that these can be included in the documentation for the construction arrangements. It is possible to take the view that one way of containing political risk is to involve as many export credit agencies as possible from different parts of the world so that, should a political problem arise, the maximum pressure can be exerted by a number of governments. On the other hand, the greater the number of export credit agencies involved in the project, the more complicated the process of finalising the arrangements is likely to be.

Collateral Warranties and Direct Agreements

In relation to the project contracts, the funder will want collateral warranties or direct agreements with the parties (including the construction contractor) contracting with the project company. One of the purposes of collateral warranties and the main purpose of direct agreements is to ensure that these agreements do not immediately fall away where the project company is in breach of its obligations under them. Where the funder enforces its security, it will want to ensure that the project company can continue the contracts under its control or that a third party acquiring the project is able to do so. Collateral warranties and direct agreements are discussed in more detail in section 7 below.

Construction Project Structures and Design Risk

Although funders are usually involved in a project after the contract structure for the construction project has been chosen by the project company, this is by no means always the case and, in any event, the funder's choice of a project can be affected by its analysis of the risks inherent in the project structure.

The most common construction project structures allocate the risk between the employer and the contractor in different ways which have been much discussed in recent years. A common thread is the relationship between the contractual (as distinct from other) risks taken by the project company on the one hand, and the degree of its control and flexibility over the carrying out of the project on the other (although a project company would argue that its own control of the project can reduce risks in relation to quality).

If a contractor agrees under a design and build contract to accept an absolute responsibility for a particular result, such as "fitness for purpose" or some other performance criterion, the project company should be able to claim for defective work more easily than under the traditional form of contract. This is because under a traditional form of contract, the design is prepared by the architect or engineer employed by the project company who owes the project company a duty of "reasonable skill and care" (rather than an absolute duty), and the contractor's duty is to comply with the detailed drawings and specifications forming that design (whether or not the design is defective).

However, the assumption of absolute liability by a contractor will usually result in the contractor requiring a greater degree of control over the carrying out of the project. In a design and build structure the project company will, therefore, have less control over the design than under the traditional form of contract and, compared with a traditional form of contract, less flexibility to make changes in it over the life of the project. Changes to the project company's requirements during the course of the project may be impracticable or expensive. In addition, unless the project company can persuade the contractor to accept a design prepared by it or its professional advisers, it will be more difficult than under a traditional form of contract to ensure value for money, adherence to detailed specifications and quality control, since the detailed design will be under the control of the contractor. Thus in a design and build structure, a project company will have less contractual risk but also less control than under a traditional contract structure.

Turning in the other direction, in a construction management structure the project company will have more contractual risk but also more control than under a traditional contract structure.

There will be the increased risk for the project company of contracting with, and co-ordinating, a large number of trade contractors. Furthermore, the liquidated damages for delay payable by the various trade contractors will usually be less than those payable by a single contractor under a traditional main contract. However, these increased risks will need to be balanced against the advantages to the project company of construction management. These advantages might include ensuring competitive tendering of the trade contracts, control over the management of the project and flexibility in the timing of the tendering, carrying out and completion of each trade contract package and in the termination of individual trade contracts. The project company will not, however, be able to pass on all the additional risks under this structure to a construction manager employed by the project company to manage the trade contracts on its behalf. This is because the construction manager, under its agreement with the project company, will only to have a duty to manage the project with reasonable skill and care, rather than an absolute duty to achieve a particular result of the type which can be undertaken by a single main contractor.

In analysing a project structure a funder will generally look for contractual risks (and therefore control), to be moved away from the project company to independent parties with the necessary resources to perform their contractual obligations. As mentioned above, a funder will also look for certainty in the obligations of those parties.

Thus, whereas a project company and its shareholders may see the increased flexibility and control over the project in a construction management structure as contributing to the general quality of the development, which may not be very easy to quantify or demonstrate, a funder will tend to focus on the increased contractual risks in terms of time and money. The increased contractual risks for the project company which a funder may perceive in construction management, compared with a design and build structure, will therefore make the structure less attractive to the funder than to the project company. From the funder's perspective, construction management may only be justified to the extent that the funder is persuaded that it is an important factor in enabling the project company to repay the loans.

On the whole, funders will therefore tend to prefer design and build, particularly for large scale infrastructure or industrial works. For commercial developments they may also favour this procurement method (at least as far as the shell and core works are concerned) but will generally accept the traditional contract structure assuming that appropriate appointments and collateral warranties are available from the professional team.

Even in design and build projects, there may be some reliance by the contractor upon information or design provided by the project company. A funder will want to identify those design elements for which the project company is responsible under the construction contract and to examine the recourse which the project company has to third parties in respect of them. For instance, in an industrial project there may be a process licence or a preliminary design supplied to the project company which is included in the contract as part of the employer's requirements and the project company should have adequate remedies against the party which supplied it. It will also be important for the funder to ascertain whether the design obligation of the contractor is an absolute duty of "fitness for purpose" or to fulfil specified performance criteria or whether it is lower standard of "reasonable skill and care". It will also be relevant whether any absolute obligations continue following the successful passing of performance tests which reflect the required standards.

Construction Contract Terms

As mentioned above, a funder is likely to look at the strength of documentation and, particularly where its time for evaluation of the project is limited, it may restrict its analysis to a number of key issues. In addition to the general matters relating to project structure and design mentioned above, these are likely to include the following:

Definition of Completion and Completion Tests

For commercial property the definition of completion entitling the contractor to require handover will need to be consistent with the requirements of any agreements with tenants or purchasers since the economic viability of the project may depend upon the letting or sale of the property.

Similarly, where industrial facilities are involved, the performance tests or completion tests under the construction contract will be vital to the funder since the project's income stream and its ability to repay the loan is likely to depend upon the facility fulfilling certain performance criteria. In addition, environmental criteria may need to be included in the tests. Failure to comply with environmental laws might give rise to fines or, in extreme cases, cause the shutdown of the plant. The funder's technical adviser is likely to have a role in certifying or reporting on the carrying out of the tests. Under some industrial construction contracts completion and handover to the project company are dependent only upon "mechanical completion". Mechanical completion occurs when the plant is physically completed and, although individual parts of the plant may have been commissioned, it takes place before the plant operates as a whole and the performance tests for the plant have taken place. The funder will prefer completion to be dependent upon the successful passing of the performance tests (or at least some of them) to acceptable, if not the guaranteed, levels. Where some or all of the performance tests need not be passed at the guaranteed levels before completion, the funder will seek to ensure that the project company has adequate remedies for shortfalls in performance, either on the initial tests or on repeat tests after remedial works have been carried out or attempted. These remedies might be by way of liquidated damages for performance to compensate for limited shortfalls in performance and equipment replacement to remedy more extensive failure.

Defects Liability

The nature of the liability of the contractor for defects, in particular the length of the defects liability period and the extent to which the contractor is responsible for latent defects after the expiry of that period, is also likely to be important to the funder. Defects liability periods of up to two years where claims are excluded after the expiry of the period may be acceptable in relation to mechanical works but are unlikely to be acceptable in relation to civil works where defects may occur several years after completion. Attention will be focussed on the relevant wording in the contract, particularly in the light of the Court of Appeal decision in *Crown Estate Commissioners v. Mowlem*.[3] The contractor will usually be able to exclude liability for consequential loss so that it is only liable for the cost of repairing the physical defects or damage.

---

[3]   *Crown Estate Commissioners v. John Mowlem and Company Limited* (1994) 10 Const LJ 311

Payment Systems

In relation to the valuations or stage payments, the funder will want to ensure that payments to the contractor, which will be funded by the funder's loans, will at all times be consistent with the value of work performed and the value received by the project company. This will generally be the case in commercial developments, particularly in the U.K.

In industrial and infrastructure projects, a common mechanism is the "milestone" system under which the contractor cannot be paid specified amounts until a particular progress stage has been reached and this or any other progress payment system will be reviewed by the funder's technical adviser. Where the funder has concerns about the value received by the project during the course of construction, it may be able to derive comfort from the level of the retentions and the terms of their release. Where retention bonds are offered in lieu, the terms of the contract and of the bonds, particularly in relation to the expiry of the bonds, will be important to the funder.

In large industrial projects where the contractor would, under the traditional valuation system for work performed on site, be paying out very large sums to sub-contractors and suppliers before payment by the project company, the contractor may seek to achieve "cash neutrality". This means that it is funded under the construction contract for payments which it in turn has to make to its sub-contractors. This will mean that the project company is drawing down under the loan agreement (and the contractor is receiving) substantial amounts in excess of the value received by the project company in terms of engineering or physical work on the ground. A funder will therefore be reluctant to accept cash neutrality, but if it does so, it will need to be confident that the project company has adequate recourse against the contractor in respect of non-performance. The adequacy of the parent company guarantee and third party bonds and guarantees will be particularly important to the funder in these circumstances. Where cash neutrality is conceded, the funder may wish to review the terms of payment to sub-contractors.

Provisions are likely to be included in the loan agreement stating that amounts cannot be drawn down until the funder's technical adviser is satisfied that the amounts are properly due under the construction contract.

Time and Money Events

A funder will want to reduce to a minimum the circumstances entitling the contractor to additional time or payment under the contract. Conventional risk management theory would indicate that, as between an employer and a contractor, risks should be borne by the party most able to control the event giving rise to the risk and that, in the case of force majeure events beyond the control of both the parties, the risk should lie with the employer. Although it might appear reasonable for the funder to accept this, the funder will be reluctant for a risk to be borne by a party, such as a project company with limited equity, which is not capable of bearing the additional cost which may arise from the risk in question. Where a risk is taken by the project company, this could result in delay and/or additional expense on the part of the project company which extends the repayment period of the loan and the funder will want to ensure that there is adequate equity, contingency funding or shareholder support or guarantees to cover the risk.

Liquidated Damages and Delay

A funder will wish the level of liquidated damages payable for delay or for shortfalls in performance to reflect, as far as practicable, the loss suffered by the project company. In practice, however, liquidated damages are widely regarded as a way of limiting the liability of the contractor for consequential loss.

Liquidated damages for delay are therefore unlikely to compensate the project company for all the loss suffered as a result of the delay. This is one of the reasons that the funder may seek a completion guarantee from the shareholders.

The funder may also be concerned to ensure that the project company has adequate remedies in the case of a prolonged delay in completion. The project company may be very unwilling to terminate the construction contract where the contractor is in breach and the funder is likely to share this view. However, when the liquidated damages for delay are in effect limited in time by being capped in some way, the only remedy available to encourage timely performance by the contractor after the liquidated damages have expired may be termination. This may be particularly relevant where a cash neutral payment system reduces the incentive to the contractor to progress the works. The rights of the project company to terminate the contract or take over the project before it has been completed may need to be carefully drafted to avoid inconsistency with the provisions for liquidated damages for delay which, to be legally effective under English law, must be the sole remedy for the breach in question.

Bonds and Guarantees

The funder will not want to take any credit risk on the contractor and so the funder will wish to assess the adequacy of any advance payment, performance, retention or other bonds or guarantees given by banks or other institutions on behalf of the contractor to support its obligations. Where the contractor is a subsidiary, the funder will require a parent company guarantee and the funder will wish to examine the creditworthiness of all parties giving bonds or guarantees.

Insurance

Insurance during the construction phase is likely to be a key issue. Ideally, insurance cover should include business interruption insurance where the project is delayed due to the occurrence of an insured risk. The loan agreement will contain detailed requirements in relation to insurance and the funder will usually wish the project insurances to be taken out by the project company rather than by the contractor since the funder, through the loan agreement, will have more control over the project company than over the contractor. Where insurances are taken out by the project company, contractors and sub-contractors who could be responsible for damage covered by the policies may require to be named as joint insureds or to have the benefit of a waiver of the subrogation rights of the insurer against them.

The funder may require insurance monies to be available to repay the loan rather than continue with the project in the event of a substantial loss. It will need to ensure that this is consistent with the insurance policy, the construction contract and any relevant local legal requirements. This requirement of the funder to take the insurance monies and not to allow reinstatement may appear unreasonable to the contractor. The funder, on the other hand, will see reinstatement as effectively a new project for which it needs the opportunity to renegotiate the financing.

Conflicts of Interest

Conflicts of interest are not unusual in project financings, and indeed are normal in BOT projects, since one or more of the shareholders in the project company may have an interest in a contract with the project company. This will often include the construction contractor and the contractor will inevitably seek to obtain the most advantageous construction contract and will not see itself as promoting the interests of the project company of which it is a shareholder. Similar considerations apply in relation to contracts with an operator or off-take purchaser where these parties are also shareholders.

As a fixed return lender, the funder will also be concerned to ensure that the project company and its shareholders cannot benefit uncommercially from the project before repayment of its debt. It will therefore wish to preserve the priority of the repayment of its debt over any return or other benefit to the equity investors in the project. Such a benefit could be given if the project company entered into a project contract so as to benefit a shareholder beyond the extent normal for a party contracting on commercial terms.

A lending funder and its advisers will therefore seek to ensure that the project company obtains a deal which is at least on arm's length terms. In addition, however, the funder may also want the project company to pass on to the contractor additional risks which the funders are not prepared to take in view of the level of equity in the project company and the sponsor guarantees or support available. Thus the funder may expect more onerous construction contract terms for the contractor and reduced risk for the project company because of the limited resources which the project company has to fund additional costs arising under the construction contract.

Where there is a conflict of interest, a funder will also wish to ensure that the contract is monitored and enforced by the project company on an entirely impartial basis. Thus the party contracting with the project company should not, as a shareholder or through a nominated director, be able to block actions by the project company in relation to claims under that contract. This could be avoided by providing, in the joint venture arrangements between the shareholders in the project company, that the directors nominated by the contracting party are not entitled to vote on the relevant resolution or by ensuring that there are independent directors who can provide a majority vote binding on the board of the project company. It will also be important to see that the shareholder concerned cannot prevent a quorate board meeting being held to deal with any matters arising under the contract. Where disputes arise, the funder will want the project company to erect "Chinese walls" to deny the construction contractor access to confidential information of the project company in relation to the dispute where it would otherwise be entitled to that information through its shareholding or nominated directors.

Collateral Warranties and direct agreements

General

Collateral warranties have become a standard feature of the requirements of funders in commercial developments, as have direct agreements in industrial and major infrastructure project financings. Each are agreements between the funder and the contractor to which the project company may also be a party.

Collateral warranties and direct agreements will vary in weight and character depending upon the type of project. For instance, a funder lending to a commercial development will tend to focus on rights in relation to defects and the assignability of the warranty to a number of potential purchasers. On the other hand, a funder of a power station project may be more concerned with the operation of "step-in" rights during the course of the contract (which would allow the funder to take over the contract or nominate third parties to do so) and the transfer of the contract to a single purchaser (whether by novation or assignment). This difference of emphasis arises because in a commercial development a funder enforcing its security may wish to dispose only of the assets of the project company, perhaps in a large number of lots, whereas in a power project the funders would be looking for a single purchaser to take over the whole project, including all the project contracts, either by a sale of the project business and assets as a going concern or possibly by a sale of shares of the project company.

Collateral Warranties

A funder requiring a collateral warranty will be particularly concerned about the nature of the liability to the funder of the party giving the warranty. This will include a separate obligation to perform the contract by the party giving the warranty in favour of the beneficiary of the warranty. The funder will look at any limitations on that liability, for instance, the exclusion of liability for consequential loss or for late completion, or limitations by reference to the liability of the contractor or professional adviser under its agreement with the project company. Ideally, the funder will not want any counter-claims by the contractor or the professional adviser against the project company (for instance, for outstanding payments) to reduce the liability of that party to the funder under the collateral warranty. Contractors and professionals are likely to resist this strongly and to argue that they should have no greater liability to the funder under the collateral warranty than they have under their agreement with the project company.

The funder will also be concerned about the terms of the exercise of its step-in rights and, in particular, may seek to limit its liability in these circumstances to pay outstanding amounts under the underlying contract to perhaps one or two valuations or stage payments. An important element in the step-in rights, particularly when the amounts drawn down under the loan are paid to the project company rather than paid directly to the contractor, may be for the funder to know when the project company is in breach of its payment obligations under the underlying contract. Whether or not the step-in right is actually exercised in the way envisaged by the terms of the collateral warranty, a limit on back payments to a contractor on the exercise of step-in rights may encourage the contractor to give notice to the funder if the project company defaults in its payment obligations under the construction contract.

Where the contract involves design or another professional discipline, the funder will also wish to be satisfied with the level and the terms of the professional indemnity insurance, for instance that it does not contain a "pay when paid" clause upheld in the Padre Island case.[1] It will also be concerned with the continuing availability of the insurance since the insurance will be written on a "claims-made" basis and may require independent confirmation from a broker, on entry into the warranty and subsequently, that the insurance is in force.

The funder will also be concerned about the assignability of the warranty to potential purchasers. As mentioned earlier, in commercial developments, the value of a collateral warranty in relation to defective work will not generally lie in the enforcement of claims by the receiver on behalf of the funder, but in the receiver's ability to assign the collateral warranty to a third

---

[1] *Firma C-Trade S.A. v. Newcastle Protection and Indemnity Association, Socony Mobil Oil Inc and Others v. West of England Shipowners Mutual Association (London) Limited (No. 2)* [1991] 2 AC 1

party purchaser of the development or part of it, or in its ability to call for further warranties in favour of tenants and purchasers once identified. One method of achieving this is to provide for one warranty to be "split" into further warranties in respect of separate parts of the development.

A major point which funders should bear in mind is that warranties from contractors in respect of developments halted due to a default by the project company are rarely of value since normally the principal duty of the contractor is generally to complete the works rather than to achieve any intermediate performance before completion. The position is to some extent different for consultants whose principal duty is likely to be a continuing duty to take reasonable skill and care throughout the term of their appointment.

Other concerns of the funder to be addressed in a collateral warranty will include access to and copyright in the project documents, in particular, plans and drawings and technical documents which may be required to facilitate the completion or sale of the development. Such access and copyright should not be dependent upon all fees having been paid since these may be the subject of a dispute.

The form of collateral warranty should not be agreed by the project company with the contractor or professional before being approved by the funder. However, the project company may need to be active in negotiating to reduce the funder's requirements, since it will need to maintain the relationship with the contractor or professional and will not wish to invite a request for additional remuneration.

Direct Agreements

Direct agreements are an important part of the funder's security package, particularly in industrial and major infrastructure project financing, enabling the funder to enforce and realise its security in the project.

The principal purpose of a direct agreement is to allow the funder and, subsequently, a third party to take over the project following the enforcement by the funder of its security. These step-in rights may need to be co-ordinated with similar rights in relation to a wide variety of other agreements and with the rights under the underlying contracts in the event that the funder needs to exercise the rights. In BOT projects (which are mentioned below) the concession arrangements will also need to allow the funder to substitute a new project company in these circumstances.

A direct agreement will allow the funder to suspend the rights of the contractor to terminate the contract for a period of time while the funder takes a view on the action required. The right of a contractor to terminate the contract would typically arise in the event of non-payment by the project company (which is also likely to be a default under the loan agreement). Within the suspension period, the funder or its nominee would have the right to step in as an additional obligor for a further period during which it would be obliged to meet the project company's obligations under the contract. The additional obligor would have the right to step out at any time and would only be liable for obligations incurred up until the step-out date. During the step-in period the funder would have the right to arrange a permanent replacement for the project company under the construction contract.

Direct agreements normally limit the ability of the contractor to terminate for force majeure, although this is usually subject to the funder paying the costs of the contractor, in addition to any costs payable under the construction contract, from the time when the contractor would otherwise have been entitled to terminate the contract for this reason.

BOT Projects

There have been numerous BOT (Build, Operate and Transfer) projects in the Far East and examples in this country include the Dartford Crossing, the second Severn bridge and the U.K. Government's DFBO (Design, Finance, Build, Operate) programme for roads currently under tender. The squeeze on government finances may well make such projects more common in the U.K. in future years.[2]

Under a BOT project, a government typically grants a concession to a "project company" under which the project company has the right to build and operate a facility, for instance, roads, tunnels, bridges, electricity, water or other public or large-scale services. The project company borrows from funders in order to finance the construction of the facility and the loans are repaid from "tariffs" paid by the government or consumers (usually members of the public) over the life of the concession. At the end of the concession period the facility may be transferred back to the government. The shareholders in the project company are likely to be the contractor, perhaps one or more of the funders lending to the project company and other equity shareholders.

In some cases, the project company may have little or no equity and the funder will be concerned to ensure, so far as possible,

---

[2] For a further discussion of BOT schemes see Scriven, J.S. *A Banking Perspective on Construction Risks in BOT Schemes* [1994] 11 ICLR 313

that all the construction obligations of the project company under the concession agreement, and those required to fulfil any performance criteria needed for the commercial success of the project, are appropriately reflected in the construction contract and, as far as possible, passed on to the contractor on a "back-to-back" basis. These "back-to-back" obligations in the construction contract could relate to provisions in the concession agreement for design, the dates for possession and completion, intellectual property, conduct of the works, suspension, termination, force majeure, warranties, indemnities and other detailed terms. The contractor, on the other hand, though it may be a shareholder in the project company, is likely to resist this effective transfer of obligations to it from the project company. It will argue that it should be contracting with the project company as an independent commercial entity which should take the risks normally taken by employers associated with construction.

The difference in perspective between the funder and the contractor may be further complicated by the fact that the government may regard its concession agreement as being in some ways similar to a construction contract, particularly since the concession may relate to a public service facility which the government will eventually own at the end of the concession period (and may do so earlier if the concession is terminated earlier on the occurrence of certain force majeure events). The government may therefore wish to retain control over a large range of construction-related matters through provisions in the concession agreement, and both the government and the funder may wish to see the obligations of the concession company uner the concession agreement passed through to the contractor on a back to back basis.

Taking this "back-to-back" principle one stage further to protect the project company and its assets, the funder may seek to limit the claims for additional payment by the contractor against the project company under the construction contract to the amounts which can be passed on as a tariff either to the government or to the consumer (perhaps by an extension of the concession period) under the concession agreement. Similarly, the funder may also want the construction contract to exclude the contractor's claims for extensions of time under the construction contract, except to the extent the project company is entitled to extend the concession period. Where the contractor is one of the shareholders in the project company, the conflicts of interest between the shareholders in the negotiation of the construction contract may become acute.

In view of the close linkage between the claims under the construction contract and under the concession agreement in BOT projects, it may be convenient for all the parties for there to be the same forum for the resolution of disputes between the government and the project company under the concession agreement as for the resolution of disputes between the project company and the contractor under the construction contract.

Conclusion

The squeeze on public funding and the relative scarcity of private equity for construction projects (together with the continuing moves in the direction of private funding of infrastructure and energy projects) seem likely to result in bank funded project financing continuing to be a major feature in the domestic and international construction scene.

In many financings in the last few years there has, however, been a change in the position that the funder's recourse in all circumstances should be totally limited to the obligations of a single purpose project company and its security limited to the assets and contracts of the project. Completion guarantees by the equity investors in the project company have become more common. In commercial developments the "buyer's market" in recent years resulted in some resistance by tenants to full repairing leases which include the structure of buildings, although this may well be a temporary phenomenon. It was the ability of a single purpose project company to let a completed building on a long-term full repairing lease and then sell its reversionary interest on to an institutional investor (which would then have a guaranteed rental income with little liability to the tenant for the structure) which made feasible much of the limited recourse financing of property development of the late nineteen eighties.

But while these developments have resulted in changes in the financing of certain types of project, there seems little doubt that project financing of construction projects is likely to continue. The financings of major power station projects in the U.K., examples being the Teeside Power project which raised £795,000,000 and the more recent Humber Power project, are evidence of this. Experiences of project failures in the property sector may only have encouraged funders to assess the construction risks in projects more carefully. Project companies and construction contractors therefore need to understand the funder's perspective on their projects and be sympathetic to the funder's concerns although these may at times seem to them to be peripheral to the success of the project. Contractors will need to understand that the construction contract may only be one part of what can be an immensely complex scheme of documentation for the project and that the funder will need to review and, in some cases, negotiate a very large number of project documents. All these documents, including the construction contract, will need to be integrated into the funder's loan and security documentation. Patience and co-operation on the part of all those involved is likely to be required.

Part II

Management and Procurement Issues

5. Impact of Risk Allocation and Equity in Construction Contracts[1]

David B.Ashley, James R.Dunlop & Michael M.Parker

Synopsis

This paper examines the ground rules for allocating risk in construction in terms of the construction contract, that the framework laid down by the construction contract defining and limiting the rights of the parties to accomplish their goals, and thus serving as a vehicle for achieving both owner and contractor objectives.

Introduction

The complexity of the contracting process cannot be overemphasized and has undergone numerous investigations to streamline its functionings. One of the more encompassing investigations derived a flow chart (see Figure 1) to help illustrate the workings of the process.[2] In its study, the Business Roundtable identified eight steps common to a typical contracting process. In order of execution, the eight steps are:

---

[1] Construction Industry Institute, Source Document 44, March 1989. This extract from the Report is reproduced with kind permission of the University of Texas at Austin

[2] *Contractual Arrangements*, A Construction Industry Cost Effectiveness Project Report, The Business Roundtable, New York, NY, October, 1982.

(1)   Execution Strategy
(2)   Contracting Strategy
(3)   Validation of contractor
(4)   Analysis and selection of any incentives
(5)   Analysis of cost liabilities or impacts of risks
(6)   Contract language
(7)   Contractor selection
(8)   Contract administration

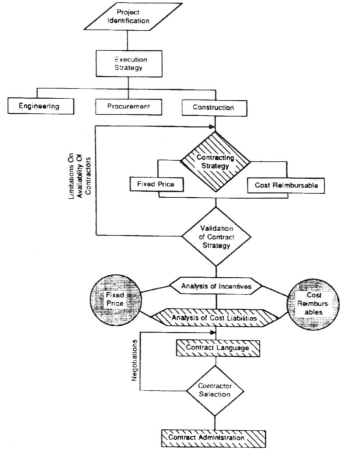

FIGURE 1 - The Contracting Process[3]

---

[3]   Source, *ibid.*

The steps of dominant concern in this study, steps 2, 5, 6 and 8, are arranged in a slightly different fashion as shown in the outline below:

1) Contracting Strategy (Step 2)
   a) Strategies Available
   b) Selecting a Strategy
   c) Trends

2) Risk Evaluation and Management
   a) Risk Management (Step 5)
   b) Contract Language (Step 6)

3) Contractual Relationship
   a) Setting the Stage
   b) Playing According to the Rules (Step 8)
   c) Possible Impacts

Contracting Strategy

Construction contracts employ a wide variety of strategies and payment terms to deal with a diversity of project complications. The two principal types of construction contracts are cost reimbursable and fixed price. The differentiating factor is cost responsibility. In a cost Reimbursable contract the owner shares in the responsibility where as in a fixed price contract the contractor had primary cost responsibility.

Strategies Available

In a cost reimbursable contract a contractor performs work for a fee plus the actual cost of the work. The basic characteristics of this type of contract most often include:

- maximum owner involvement and ability to analyze costs;
- incomplete or unclear work scope definition and cost;
- minimization of overall design and construction time period; and
- importance placed on owner-contractor relationship and trust.

In a cost reimbursable contract, the contractor is always reimbursed for direct costs and variations only occur in indirect costs and fee coverage. Variations of the cost reimbursable contract often encountered include:

- Percent Fee: Contractor reimbursed for costs and a percentage of costs.

- Fixed Fee: Contractor reimbursed for costs plus stipulated sum covering general administrative costs and profit.

- Incentive Fee: Contractor reimbursed for costs. Fee dependent upon achieving certain costs or schedule goals.

- Performance Fee: Fee varies according to a certain agreed upon criteria on which the contractor is rated for performance.

- Conversion: Any type of reimbursable contract converted to a fixed price or guaranteed maximum contract.

The advantages of a cost reimbursable contract depend upon the particulars of a project, but general advantages from an owner's stand point were clear. First and foremost is the time advantage. A cost reimbursable contract can be run in a fast-track manner, phasing and integrating design with construction and greatly reducing project length. Second, cost reimbursable contracts reduce the adversity typically imposed by a contractor striving for profit while an owner tries to minimize cost. Third, this system allows great flexibility for the inevitable changes that occur on every project and that occur more frequently on cost reimbursable projects. Fourth, the owner can ordinarily count on a higher standard of care and better quality with a "pay as you go" scheme. Lastly, where contractors are unwilling to accept high risks or where project scope cannot be nailed down, cost reimbursable contracts offer an alternative that most contractors were willing to accept.

Many of the disadvantages of cost reimbursable contracts directly follow from the advantages. First, cost reimbursable construction generally requires a great deal of owner involvement and, consequently, risk assumption (including responsibility for costs). Second, the final cost of the project is not guaranteed. Third, cost reimbursable contracts tend to be less economical than fixed price. Lastly, cost reimbursable contracts require a more extensive contractor selection and negotiation process than fixed price contracts.

A fixed price contract was one in which the contract agrees to perform all the work specified for a stipulated sum of money, regardless of the cost.[4] The point to be emphasized is that the contractor agreed to the stipulated sum regardless of trouble and expense encountered. This highlights the points that if a fair and reasonable price cannot be established or if the scope, design and specifications are not sufficiently detailed or clear at the time of bidding, then a fixed prices contract should not be used. Some of the variations of the fixed price contract include:[5]

- Lump Sum: "Hard money contract" - Single fixed price for entire contract.

- Unit Price: Payment on the basis of units of work actually done. Unit costs and estimated quantities stated.

- Fixed Price with escalation: Price adjustments on cost of certain materials, labor or other factors beyond contractor's control.

- Fixed Price with Bonus/Penalty for completion schedule: Amount per day bonus/penalty for early/late completion.

- Guaranteed Maximum: Price ceiling; bonus/penalty for cost under-runs/over-runs.

The benefits derived by an owner from fixed price contracts are many and as usual depend on project type, owner sophistication, and owner risk attitude. The biggest advantage is having the general, overall cost determined and cost responsibility shifted to the contractor. For owners without supervision capabilities, this

---

[4]   Dunham, C.W., Young, R.D. & Bockrath, J.T., *Contracts, Specifications and Law for Engineers*, McGraw-Hill, Inc, 1979.

[5]   *Contractual Arrangements, Op cit*

option minimizes owner involvement. The owner benefits from price competition and the contractor has significant incentives to meet schedule, reduce costs, and improve productivity. One last advantage is predictability - fixed price contracts have well established legal and contractual precedents.

Other than a lengthy design-construct time, most of the negative influences of fixed price contracts stemmed from the fact that the contractor bears the economic risk of many factors not under contractor control. These economic pressures place the owner and contractor into adversarial roles. These forces also minimize a contractor's design constructability input and motivation to improve quality. A contractor might neglect to mention necessary changes for fear of non-reimbursement. Lastly, contract financial difficulties, caused by an insufficient by an, can cause serious delay problems to an owner who simply desires a completed project.

Strategy Selection

Selecting a contracting strategy becomes a question of merging the most favorable aspects of the two contract types to fit the project goals. The trade-offs in this merger involved and depend upon the risks assumed in relation to the overall project goals. "The three objectives of cost, time, and quality must be analyzed and placed in some priority, since trade-offs will probably be necessary in deciding what type of contract is to be used".[6] Thus, the owner's ranking or prioritizing of the three principal goals will determine the contract type.

Some figures presented more clearly illustrate given trade-offs corresponding to contract types. Figure 2 displays disparities in project duration related to contract type. Figure 3 reveals

---

[6] *Contractual Arrangements, Op cit*

differences in owner risk, owner control, and information required.

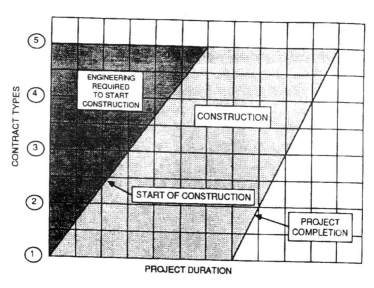

Figure 2 - Project Schedule vs. Type of Contract

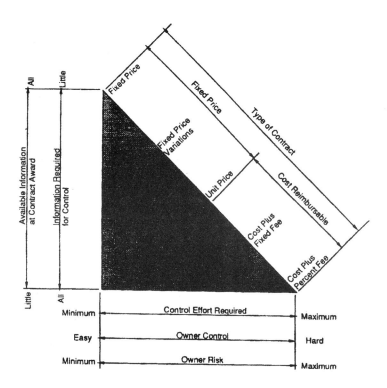

Figure 3 - Factors Influencing Contract Choice[7]

[7] Barnes, M.L., *Advanced Construction Project Management Techniques*, Continuing Engineering Studies Seminar, Civil Engineering Department, The University of Texas at Austin, November 2 & 3, 1983.

Contractor goals should be considered in addition to the owner's goals. Since a contractor will react to personal objectives based on contract type, only an unseasoned owner would fail to consider contractor motives. For example, market pressure may drive a contractor to accept high risks with a low profit potential based on a simple instinct to survive. Yet, is it advisable to place the contractor in a situation where financial troubles may subject an owner's project to multiple disputes, great delays, or shoddy work? "The owner's goal of winning a commitment must always be weighted against the risk of not achieving one or more of the overall project objectives".[8] Consequently, a wise owner will consider contractor reactions in order to maximize potential returns.

Trends

Recent trends in the construction industry have created a "buyers market" for construction services. Often times owners are able to use their bargaining power to dictate not only contract type, but also specific contract language. Owners go unopposed because contractors are willing to accept these high risks rather an go without work. Some owners use their power to obtain fixed price contracts and then attempt to administer them as cost reimbursable contracts. Not only does this enhance an adversarial relationship, but many times in the event of trouble, the courts examine the actions of the parties to determine contractual intent, throwing out any onerous clauses. Hence, some owners are paying fixed price contract premiums and then resuming, through their actions, the liability they sought to shift. In any case, the action conflicts with good business sense.

---

[8] *Contractual Arrangements, Op cit*

Risk Evaluation and Management

Risk taking and thus risk managing have become notable areas of concern in recent years due to demanding economic times and the intense competition in the construction industry. The construction contract is the focal point for the assignment of risks inherent in a construction project. Risks and responsibilities associated with a specific project must be clearly allocated within the contract. Although the owner may decide the allocation policy for a project it is the responsibility of both parties to provide adequate risk management.

Risk Management

Complex, unpredictable and uncertain risks require the pragmatic, systematic treatment of a risk management procedure. "Risk management is the art and science of identifying, analyzing and responding to risk factors throughout the life of a project and in the best interest of its objectives".[9] The risk factors comprise the characteristics of a risk such as the risk event, its probability of occurrence, and the amount of potential loss or gain. Risk management typically involves the following functions: risk identification, impact analysis, response planning, the response system, and application of the resulting date. Three broad categories will be discussed here: identification, analysis, and response.

Risk identification is an essential first step for a successful project. Since owners draft contracts and have ultimate responsibility for project cost, they are responsible for making a thorough evaluation of all potential risks and responsibilities including: adequacy of design, cost of construction, liability to subcontractors, indemnification for all casualties, financing, and

---

[9] Wideman, R.M., *Risk Management*, Project Management Journal, September 1986, pp 20-26.

coordination of the work. At the earliest stages of a project, risk identification can aid the owner and the contractor in establishing project constraints. It should be evident that risk and its occurrence influence project cost, schedule, and quality, but the impacts can be controlled to the extent the risks are effectively identified and managed. A risk, however, cannot be managed unless it is properly identified.

Risk analysis has had limited use in the construction industry outside of large projects.[10] The idea of risk analysis is to quantify the impact a risk will have on the project cost, schedule, and/or quality. This is often accomplished by determining the impact a risk may have within a range of values and then assessing the probability of occurrence of the risk. There are many sophisticated techniques available for the analysis of project risks, the discussion of which is beyond the scope of this paper. The necessary information is generally obtained through either historical data files or the expertise of experienced personnel within the company. Once this information is obtained, a sensitivity analysis may be performed and the resulting expected risk impact realized. It should be mentioned that isolated independent risk occurrences are not the norm. Influence diagrams and interdependency.

Risk response is a process for formulating risk management strategies - mitigation, deflection, and contingency planning.[11] Mitigation, in essence, means to lessen the effect of a risk. Many times, an owner must reappraised the project execution as a whole once it is determined that mitigation is necessary. Identification of excessive risk may warrant a change in scope, budget, schedule, or quality in order to better manage high impact risks. It may be possible to reduce a risk's potential by

---

[10] Perry, J.G. & Hayes, R.W., *Risk and its management in construction projects*, Proceedings of the Institute of Civil Engineering, Part I:499-521, June, 1985.

[11] *Risk Management, Op cit*

redesign, different contracting strategy, or different construction methods.[12]

Deflection or transfer of project risks is a common response to construction risks. A transfer can be a total allocation to another party or a risk-sharing between two or more parties. This may be done through the purchase of insurance or by contracting the responsibility to the other party. It is often advantageous for an owner to shift risks to the contractor, but it must be understood that there is a premium associated with this shift. If the owner is going to pay for this type of protection there must be a careful examination as to whether it will effectively accomplish the project goals. Questions to be answered during this examination are:[13]

- Which party can best control the events which lead to the risk occurring?
- Which party can best control the risk if it does occur?
- Do you want to have any involvement in the control of the risk?
- Will the premium charged by the party to which the risk is allocated be reasonable and acceptable?
- Will this party have the ability to bear the consequences of the risk should it occur?
- Will the allocation of this risk lead to other risks being transferred back?

The problem is that "as long as the costs associated with their (the owner) shedding of risks remain obscure, ill-defined, and unquantified, owners will naturally adopt those strategies designed to minimize their susceptibility to variations in costs,

---

[12]   Perry, J.G. & Hayes, R.W., Op cit
[13]   Perry, J.G. & Hayes, R.W., Op cit

and contractors and designers will naturally charge premiums for their increased susceptibility to these risks".[14]

The premiums charged by contractors are in the form of contingency funds included in their bid. A percentage of the total price is added to the bid as a cushion to pad the contractor's profit in the event of the unexpected.

General Liability Insurance is commonly required of the contractor on a construction project. This insurance protects the contractor from third party lawsuits claiming that the contractor did not act in accordance with tort law. The availability and affordability of business insurance for contractors has been a problem in recent years with increases ranging from 200% to 700%.[15] Reasons for this dramatic increase in insurance premiums are listed below:[16]

- Expanded liability for professionals;
- Poor underwriting by insurers;
- Decline in stock market which reduces the value of insurance company investments;
- Downturn in economy; and
- High cost of defending claims.

The effect of this increase in cost and decrease in affordability is an increase in project cost. If insurance is available, the higher premiums will directly increase project costs. If insurance is not available, the contractor will normally include a large

---

[14] Levitt, R.E., Ashley, D.B. & Logcher, R.D., *Allocating Risk and Incentive in Construction,* (1980) 106 Journal of the Construction Division, ASCE, 297-305, September, 1980.

[15] Pitts, H., *Prospective Contractor Bondability Seminar,* Bill Pitts Insurance Agency, Austin, TX, September 17, 1986.

[16] Sweet, J., *Legal Aspects of Architecture, Engineering and the Construction Process,* West Publishing Company, New York, NY, 1985.

contingency to assume the risk of self insurance. Finally, if the contractor does not provide for the risk at all, the owner assumes the ultimate risk of an bankrupt contractor and associated delay in finding a new contractor in the middle of a project.

Contract Language

Owner risk allocation is achieved through language used in the contract provisions. Exculpatory clauses are often used to shift one party's common law responsibility to another within construction contracts. These clauses are very controversial in nature because they are contrary to the common law principle that each party should be responsible for its own wrongful acts.

There may be agreement that an exculpatory clause which shifts all risk to the contractor is unconscionable, but the real issue in today's construction industry seems to be a question of what the courts will allow - what can we legally get away with? Although it is perfectly legal for parties to contract and bind themselves to an exculpatory clause, the courts tend to look at the relative bargaining positions of the parties and try to find an equitable result rather than enforce exculpatory clauses.[17] Due to this lack of strict legal precedent, courts find many exceptions to these clauses and are therefore relatively inconsistent in their decisions. No damage for delay clauses, difference conditions disclaimers, and indemnification clauses are examples of exculpatory clauses used in construction contracts today.

Although it may seem advantageous for the owner to shift its responsibilities to the contractor, one must question its cost effectiveness. Granted, an exculpatory clause will make the contractor responsible for project risks, but most contractors merely pass this cost back to the owner in the bid price.

---

[17] Vansant, R.E., *Exculpatory Clauses: An Ineffective Technique*, (1985) 38 The Construction Specifier, pp. 17-18, March, 1985.

Therefore, by including such language in a contract and considering the court's inconsistent judgment in this area, the owner is paying the price for protection and not necessarily receiving it.

When a contract is litigated, the courts determine who is responsible under the terms of the contract. Parties to a contract have the right to conduct their business and contract in any form they choose - the courts do not have the power to rewrite the contract. However, when there are problems on a project which lead the parties to litigation, it is the duty of the courts to interpret each party's intentions. The first method used by the courts to do this is the "plain meaning rule." This rule is executed literally as a means of looking at the language used to discover the intentions or "plain meaning" of the parties. The use of this rule is rarely enough to conclude the intentions of the parties, so relevant surrounding facts and circumstances and other rules of interpretation are generally needed.[18] Not only will courts give heavy weight to the setting in which the contract was made, but will also consider the actions of the parties after the fact. For example, if the actions of one party are contrary to the contract and the other party accepts these actions, the courts will not usually uphold the strict language of the contract. Some of the rules of the contract interpretation used by courts are listed in Table 1.[19]

---

[18] *Legal Aspects of Architecture, Engineering and the Construction Process, Op cit*

[19] Levin, P., *Claims and Changes*, Silver Spring, Maryland, Construction Industry Press, 1981.

Rules of Contract Interpretation
1. The ordinary meaning of the language is given to the words unless circumstances show that a different meaning is applicable.
2. Technical Terms and words of art are given their technical meaning unless the context indicates a different meaning.
3. A writing is interpreted as a whole and all writings forming part of the same transaction are interpreted together.
4. All circumstances accompanying the transaction may be taken into consideration, except that when the parties adopt a written statement of their agreement, oral statements of their intentions concerning the agreement made prior to, or simultaneously with, the writing may not generally be considered.
5. If the conduct of the parties subsequent to a representation of intention indicates that all the parties are placed a particular interpretation upon it, that meaning is adopted.
6. An interpretation which gives a reasonable, lawful and effective meaning to all representations of intention is preferred to an interpretation which leaves a part of such representations unreasonable, unlawful, or of no effect.
7. The principal apparent purpose of the parties is given great weight in determining the meaning to be given to representations of intention or to any part thereof.
8. Where there is an inconsistency between general provisions and specific provisions, the specific provisions control.
9. Where words or other representations of intention bear more than one reasonable meaning, they are interpreted strongly against the party from which they came.
10. Where written provisions are inconsistent with printed provisions, the written provisions control.
11. Where a public interest is after an interpretation is preferred which favors the public.

Table 1: Rules of Contract Interpretation[20]

---

[20] *Claims and Changes, Op cit*

Contractual Relationship

"The language of individual contract clauses merits careful attention, but of paramount importance is the total agreement between the parties - including the total contract, external supporting documents, project coordinating procedures and less formal understanding".[21] Two of the four important items described by the previous CII study can be directly linked to the contractual relationship - the less formal understanding and the project coordinating procedures.

Setting the Stage

First consider the less formal relationship. At this stage in the relationship development, the initial stage, the parties are orally defining their capabilities, roles and responsibilities. The parties are in fact "setting the stage" for the "big production" they will undertake. Both coordination and trust drive the project while misunderstandings, concealment and gross mis-allocations of risk stifle the teamwork attitude with a "what's in it for me?" feeling. Coordination, trust and communication provide the foundation for a favorable initial relationship. The Business Roundtable Report[22] supports the establishment of good communications with a contractor in order to assess the contractor's experience and capability to assume and control risks.

In the contemplation of a contract, both parties should examine the goals and capabilities of the other party. Sweet[23] defines some of the elements of a successful contract relationship in Table 2. Certainly an oral arrangement comprising the highest

---

21 *Impacts of Various Construction Contract Types and Clauses on Project Performance*, Construction Industry Institute, Publication 5-1, University of Texas at Austin, July, 1986.
22 *Contractual Arrangements, Op cit*
23 *Legal Aspects of Architecture, Engineering and the Construction Process, Op cit*

level of each of these elements would constitute an ideal relationship and a written agreement would be unnecessary. In fact, Ashley and Matthews[24] note that"....many [of these factors] are beyond contractual definition, whose presence or absence can affect the quality of the personal interactions and ultimate project outcome." In such a case a contract simply serves as a documented remembrance of the actual meeting of the minds. Yet, it is this remembrance that creates the framework for all future interactions.

Element of a Successful Contract Relationship	
Amenability	Initiative
Competence	Managerial Ability
Congeniality	Open Mindedness
Cooperative Tendencies	Organizational Ability
Discretion	Planning Ability
Equitable Adjustment	Punctuality
Financial Responsibility	Satisfactory Performance
Good Reputation	Technical Knowledge
Interpretive Understanding	Timely Compliance

Table 2: Elements of a Successful Contract Relationship[25]

Figure 4 diagrams the fluctuation in the owner-contractor relationship over the course of a project. The diagram graphically depicts the high and lows of the owner-contractor working relationship and shows the points in time when project events influence this relationship. The initial relationship defines the

---

[24] Ashley, D.B. & Matthews, J.J., *Analysis of Construciton Contractr Change Clauses*, Vol I & II, The University of Texas at Austin, December, 1985.

[25] *Legal Aspects of Architecture, Engineering and the Construction Process, Op cit*

starting point in time of a "quality and character" of relationship line. Figure 4 shows that this starting point can be highly complementary, highly adversarial or so extremely close to the dividing line that the smallest of incidents would push it over that line. Events occurring during both the contractual arrangements stage and the contract period will influence the quality and character line.

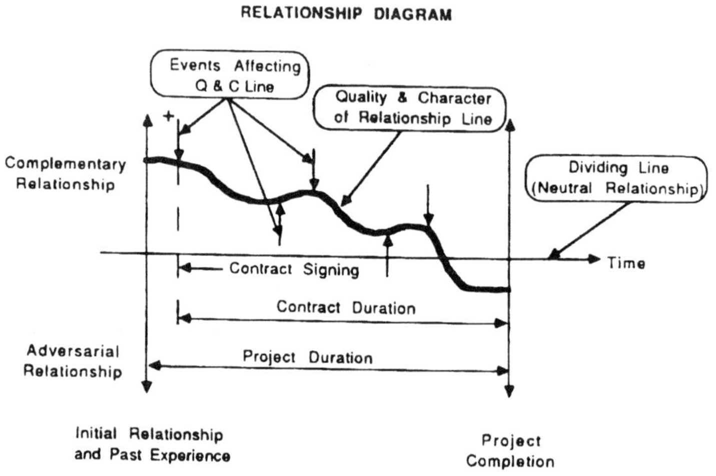

RELATIONSHIP DIAGRAM

Events Affecting Q & C Line

Quality & Character of Relationship Line

Complementary Relationship

Dividing Line (Neutral Relationship)

Contract Signing

Time

Contract Duration

Adversarial Relationship

Project Duration

Initial Relationship and Past Experience

Project Completion

Figure 4: Relationship Diagram

One can visualize interesting scenarios from examination of the relationship diagram. Consider an initially adversarial relationship: What would drive the parties to contract under these conditions? Is it market pressure or economic duress? Would favorable contract terms cause the quality and character line to cross over dividing line? Would low bid boundaries (for public owners) cause the dividing line to cross over?

Economic duress exists where a great disparity in bargaining power is used to pressure a weaker participant into an agreement. Unlike market pressure where the parties' bargaining positions are relatively equal, economic duress allows the stronger party to stipulate specific contract language and assign unfavorable risks to the weaker bargaining party. In lean construction times, owners may take advantage of contractors' economic duress, in booming construction periods contractors may apply similar "screws" to owners. True economic duress is actionable under contract law (one can file a lawsuit claiming that economic duress caused an unfavorable decision or outcome), but courts have only found true duress to exist only in a limited number of cases. Most of the previous discussion might simply connote hard bargaining in the eyes of some construction professionals. The Business Roundtable[26] provides some "words to the wise" on this topic:

> *Owners should avoid dictating preferential contract language through superior bargaining power. This creates an adversary relationship between the owner and contractor, a poisoned atmosphere that can jeopardize the owner's objectives for the project.*

Favorable contract terms might persuade a hesitant contract to jump at a job with questionable initial relationship characteristics. When the terms of a contract seem fair and complete, owners and contractors assume that they can pursue a project based on

---

[26] *Contractual Arrangements, Op cit*

the law stipulated in the contract. They plan on using the contract whenever communications or negotiations breakdown. Recalling Figure 4, if a relationship has not crumbled too far down on the relationship diagram, optimistic contractors will pursue work with a favorable contractual framework. Thus, an advantageous contract might persuade owners and contractors sign an agreement, even though their relationship is adversarial.

As a rule, governments offer construction work on a competitive, lump sum basis and by law must honor the low, responsible bid. This situation presents difficulties when a public owner has previously worked with an antagonistic, unfaithful contractor and the contractors bids low on the owner's present project. This owner should be prepared for all avenues of dispute by thoroughly drafting the contract documents in hopes of replacing much of the need for an amiable relationship.

Now, consider an initially positive or complementary relationship. Under this scheme, if favorable contract language exists and the price is agreeable, then contractors and owners rush to complete the bargain. A highly positive initial relationship burdened by a number of negative events or a crucial negative occurrence poses the only unlikely possibility of change to a negative relationship. A deal delivering a relationship just barely above the dividing line meets with favor. A project with a well-written contract and a quality an character line just at the neutral line would be acceptable and possibly favorable. However, while both parties feel the project benefits them, even the slightest hint of trouble will slide the relationship below the acceptable, below the dividing line.

Playing According to the Rules

The second task at hand involves coordinating project procedures. Contract documents define work scope, assign responsibilities, and establish the terms and conditions of the agreement. A contract functions as a tool for anticipating, interpreting, and resolving difficulties and disagreements. Contract documents define the rules and contract administrators make sure everyone is playing according to the rules.

Often misunderstood is the vital role that a contract plays in the allocation of risks. Risk allocation, the definition and division of responsibility associated with a possible future loss or gain, seeks to assign responsibility for a variety of hypothetical circumstances should a project not proceed as planned.[27] In the words of an attorney, "the contract documents will serve as the law between the parties and establish which party has assumed the risk or negated a particular risk in connection with the project".[28] The ideal contract, it is said,[29] assigns risks to the party "best equipped to manage and minimize" those risks. Management and minimization of risks occur before a contract is signed; yet for proper control of a project, contract administrators partake in pre-contract negotiations in order to acquaint themselves with the risks they must manage and to develop risk minimization strategies. Contract administration mimics risk management in all aspects except one: equity. Risk management defines, assesses the impact of, and distributes risk. Contract administration requires the human element of equity and air play in addition to the risk management elements. According to Sweet,[30] increasingly the law finds that all contracting parties owe each other the responsibility of good faith and fair dealing. In their description of the attributes of

---

[27] *Impacts of Various Construction Contract Types and Clauses on Project Performance, Op cit*

[28] *The Construction Owner, Developer, Architect and Engineer Claims: Practical Approaches to Claims Prevention,* Seminar Proceedings, Engineering News Record, New Orleans, LA, March, 1985.

[29] *Impacts of Various Construction Contract Types and Clauses on Project Performance, Op cit*

[30] *Legal Aspects of Architecture, Engineering and the Construction Process, Op cit*

proper contract administration (see Table 3, The Business Roundtable notes "equitable but firm enforcement of contract terms." Barrie and Paulson[31] describe contract administration as the "…. application of business like common sense and fair play in keeping with the terms and conditions of the contract." Clearly, administration of contracts involves the use of judgment in the interpretation of contract language.

Attributes of Proper Contract Administration
1. Complete knowledge of the contract
2. Equitable but firm enforcement of contract terms.
3. Procedures for administration established as soon as the contract is awarded and followed consistently.
4. Contract administrators at all levels knowing their authority and making decisions consistent with such authority.
5. Changes, claims and disputes resolved promptly at the longest level possible consistent with delegated authority.
6. Excessive bargaining, crisis management and threats to terminate the contract generally avoided.

Table 3: Attributes of Proper Contract Administration[32]

Thus, a contract administrators uses judgment when wielding a contractual tool in protection of his or her employer as a knight would wield a sword in defense of a king. Contract administrators, like knights, seek justice and equity in their dealings with others, but, bound by a duty to their employer,

---

[31] Barrie, D.S. & Paulson, B.C.Jr., *Professional Construction Management*, McGraw-Hill, Inc., 1984.
[32] *Contractual Arrangements, Op cit*

must also aspire to use their tools in manners that most benefit their masters. It is this juxtaposition of duties that has lead to the increasingly costly legal and insurance environment seen today.

Possible Impacts

Minimizing the impacts of a incongruent contract relationship requires a reassessment of values in terms of equitable arrangements and risk allocation before contract signing and in terms of just dealing during the construction process. The current, enormous increase in construction litigation suggest gargantuan problems in the future. This study approaches the problems encountered in the contractual relations of owners and contractors.

6. Quality Management and Risk

John Barber

Synopsis

This paper examines the relevance of quality management systems to risk in construction, and also describes progress and the impact of developments since the publication of Quality Management in Construction - Contractual Aspects,[1] notably the recent publication of the Latham Report[2] and the 1994 Edition of the BS EN ISO 9000 series.

Introduction

This paper is the fourth of what has emerged as a series, published by the Centre of Construction Law and Management, arising out of, or following on from, a Research Project undertaken by the author for CIRIA under the title 'Quality Management in Construction - Contractual Aspects'. The Research Project was largely funded by the Department of the Environment.

The first paper, 'Quality Management - Remedy or Sanction', delivered at the third CCLM Conference in 1990 and published

[1]   Barber, J.N., *Quality Management in Construction - Contractual Aspects*, SP84, CIRIA, London, 1992.
[2]   Latham, Sir Michael, *Constructing The Team, Final Report of the Government / Industry Review of Procurement and Contractual Arrangements in the Construction Industry*, HMSO, London, 1994.

in the Red Book,[3] Reported the legal background affecting quality management systems. The second, 'Quality Management - The Way Forward' was delivered as the 1992 CCLM lecture to the Society of Construction Law, and provided a summary of the main points of the CIRIA Report, SP84, which resulted from the Research Project. The third, delivered at the sixth CCLM Conference in 1993 and published in the Green Book,[4] dealt with the related topic of Environmental Management Systems, the new British Standard BS 7750 and the Eco-management and Audit Regulation.

The publication of SP84 by CIRIA in 1992 stimulated a considerable amount of interest and comment. Most of this was constructive, although some members of the technical press sought to use it to drive a wedge between CIRIA and BSI. It is possible at this stage, 2 years on, to identify which recommendations or criticisms in the Report have been matched by action, but it would be presumptuous to assert that changes or further recommendations which effectively respond to criticisms or recommendations in the Report are solely, immediately or consciously the result of the Report. Ideas tend to find acceptance when their time has come. Many of the recommendations were not, and did not pretend to be, new - they had emerged from the consultation process. Nevertheless, the experience of the Project confirms that the imprimatur of CIRIA carries considerable weight and guarantees a sizeable readership for the fruits of a research project. In addition, the CIRIA system of setting up a steering group for each project not only provides valuable contributions of ideas and guidance, but also ensures that a good number of influential persons are involved and informed, to spread the word in other forums.

---

[3]   Uff, J. & Lavers, A. (Eds), *Legal Obligations in Construction*, Centre of Construction Law, King's College London, 1992.

[4]   Uff, J, Garthwaite, H. & Barber, J.N., *Construction Law and the Environment*, Centre of Construction Law, King's College London, 1994.

Latham Report

One of the most enduring images of the 20th century is the Newsreel footage of an event in September 1938: Neville Chamberlain standing on the steps of an aeroplane, waving a document and declaring "Peace with Honour, Peace in our time". The history books record that when he returned to Downing Street, he issued a statement that:

> *"We regard the agreement signed last night and the Anglo-German Naval Agreement as symbolic of the desire of our two peoples never to go to war with one another again. We are resolved that the method of consultation shall be the method adopted to deal with any other questions that may concern our two countries."*

When we see a particular Form of Contract held out as bringing peace in our time, the words have a certain familiarity. One would not deny the importance of the content of an agreement: it may be conducive to peaceful co-operation or co-existence, or it may provoke hostility. But the lesson is surely that Agreements must be backed by force or by the true goodwill of the parties. Otherwise, they provide little or more than a breathing space before the hostilities. What has that to do with this paper on quality management and risk? The point is that Sir Michael Latham accepts the promotional hype of the NEC at face value, but he is ultra-cautious regarding the value of quality management. He requires proof that quality management works before accepting it as part of his scheme. My view is that quality management has much more to offer in promoting co-operation than he allows.

What I find encouraging about the Latham Report nevertheless is that, although Sir Michael may not have realised it, much of his underlying policy is consistent with a quality management approach, and a number of his recommendations would actually overcome current obstacles to the successful adoption of quality management. For example:

- win-win approach: quality management creates as a common goal the achievement of reliable quality in product supplied;
- recommendation of a standardised basis of prequalification: this is needed with quality management prequalification to avoid waste; and
- minimising aggregate risk: this is the key to quality management.

Above all, Sir Michael acknowledges the essential contribution of clients to the successful outcome of projects. In his words, "Implementation begins with clients".

The six paragraphs under the heading 'Quality Assurance' in the Report are to be found at 7.41 to 7.46. There is endorsement for the Construction Quality Forum, but the rest is mainly taken up with despondent comments on quality management. There is a little more enthusiasm for TQM, but the conclusion in 7.46 is subdued. It states:

"Quality Assurance certification should continue to be encouraged within the construction industry as a potentially useful tool for improving corporate management systems. But more evidence is needed that it will also raise standards of site performance and project delivery before it should be made a qualification condition for consideration for public sector work. The implementation stage of this Report should try to achieve a consensus from the industry and professions as to how BS 5750 accreditation can improve project delivery and site performance as well as office management systems. Encouraging a Total Quality approach should pervade the

whole implementation phase, it should involve heavy emphasis upon teamwork and co-operation. "

Elsewhere in the Report, however, the philosophy often appears identical to that advocated by quality management thinkers. Even more encouraging, the Report makes several recommendations or comments which answer concerns in SP84. For example, the statement in para 1.11 that "Implementation begins with clients" is followed in para 3.5 with endorsement of a statement in the DoE's Guidance Manual as to the client's proper role in controlling interfaces. This states:

"The Project Manager's contract with the client will have required the PM to install procedures to be followed by support consultants and control staff concerned with the project. "

Similarly in para. 4.1, the Report states:

"Effective management of the design process is crucial for the success of the project. It should involve... the co-ordination of the consultants, including an interlocking matrix of their appointment documents which should also have a clear relationship with the construction contract documentation. "

Perhaps the most important comment is in para. 3.7:

"The next step should be the use of internal risk assessment to devise a contract strategy. The client should decide how much risk to accept. No construction project is free from risk. Risk can be managed, minimised, shared, transferred or accepted. It cannot be ignored. "

The Report also deals with the vexed question of standardising prequalification arrangements to avoid the waste of resources in responding to individualised enquiries. This problem has become particularly serious since the requirement of quality systems and in some cases external auditing of quality systems became a regular part of the prequalification process. Para. 6.24 recommends:

> *"As a first step, and as a matter of urgency, the DOE should set up a task force drawn from the public sector as a whole to prepare a single qualification document for contractors seeking to do work for any public sector body."*

Finally, one might note the Report's endorsement of Latent Defects Insurance and Co-ordinated Project Information, and the recommendation that the Government should adopt the majority report of the Working Party on Liability. All these were supported in SP84 as compatible with and helpful to a quality management approach.

Contractual and QM Approaches Distinguished

The difference between a contractual and a quality management approach is that quality management is about the management of activities or processes. It is about doing rather than promising to do. The point can be illustrated by an example. Some years ago, before quality management had taken hold, a junior Minister complained about claims on Public Works contracts and blamed the independence of the Engineer. I wrote suggesting that his complaints were misguided and that, if he wished to reduce claims, attention should be directed to eliminating gaps and inconsistencies in contract documents. I suggested that his Department should adopt a practice of independent checking of contract documents for completeness and consistency before they were issued for tender, that is, I recommend doing something.

The reply from the Department rejected the suggestion of independent checking, with the statement:

> *"The Engineer is now required to certify that documentation accurately represents the work intended and that Bills of Quantities are a true measurement of the work shown in the tender documents."*

That is a contractual approach, which avoids having to do anything. My own experience over the past 5 years indicates that this contractual approach to eliminating gaps and inconsistencies has not been effective. Some people may well be grateful, since gaps and inconsistencies are the seeds of disputes, but I am pleased to note, in the public interest, that the same Department has since moved towards implementing or requiring quality management systems at all stages of projects.

Developments in Quality Management

I am not aware of any major developments in quality management systems since the publication of SP84 in 1992. Although the low level of construction activity during the intervening period ought to have provided a good opportunity for companies to implement or refine quality management systems, the shortage of cash has probably inhibited the investment required. There has been publicised resistance amongst smaller companies, possibly rightly so. Nevertheless, in a recent article in New Civil Engineer, David Neal, Chief Executive of May Gurney, was reported as praising the benefits obtained in his company from implementing a quality management system over the past 2 years. He is quoted as saying:

> *"Implementing the system had two advantages. A rigorous Report of all management systems found unnecessary activities and proved to be a useful 'cleansing operation'. Introducing the reporting of mistakes has changed the convention of always covering up on site. For the first time top management sees the cost of mistakes and can do someting about it."*

My basic conclusion remains unchanged, that the Model Standards in the ISO 9000 series are superbly drafted documents, which provide a comprehensive approach to quality management, but that, despite the recent revisions, they are still not entirely satisfactory for the construction sector and require tailoring.

ISO 9000 Revisions

SP84 was based on the 1987 edition of BS 5750: Quality Systems, which was self-stated to be identical with the International Standard ISO 9000 series. The ISO 9000 series was in turn the result of adoption and modification by the International Standards body of the original 1979 edition of BS 5750. ISO 9000 was also adopted by the European Standards body under the number EN 29000, and endorsed by an EC Council resolution as the approved basis for requiring quality assurance in procurement within the European Community.

Seven years on, a new edition of the five core ISO 9000 Standards has been published. One by-product is that BSI has given up the famous Standard number BS 5750, at least for the Internationally agreed Standards. The revised Standards are known as the BS EN ISO 9000 series or 'family'. The particular Standards published in the new edition are two Guideline documents, ISO 9000-1 and 9004-1, and three Model Standards, ISO 9001, 9002 and 9003. Other current or planned Standards considered to form part of the family are listed in ISO 9000-1.

It is convenient to Report the changes in the two groups of documents, the Model Standards and the Guidelines, separately, but there are three changes in terminology and definition which affect both groups. The first is the definition of 'quality', which now becomes:

> *"Total of characteristics of an **entity** that bear on its ability to satisfy stated **and** implied needs".*

The word 'features' has disappeared. The new word 'entity' replaces 'product or service'; this hints at Gallic influence, since the word is more commonly used in such a context in French than in English. The conjunctive 'and' replaces the disjunctive 'or'.

The actual and/or intended consequences of these changes are not entirely clear, but there are instances where advantage has been taken in the drafting. For example, ISO 9000-1 para. 4.1 refers to 'quality of its own operations', ISO 9004-1 section 10 refers to 'quality of processes'. These phrases are meaningful with the new definition, but would have been incompatible with the old one.

The second change is that the term 'purchaser' is replaced throughout by 'customer'. The third is a Note to the definition of 'product' which adds that:

> *"Product can be either intended (e.g. offering to customers) or unintended (e.g. pollutant or unwanted effects)."*

The addition of unintended product in the definition only applies, however, for the purposes of the Guidance documents, not the Model Standards. The new addition has obvious implications in relation to environmental aspects. The use of different versions of the definition is likely to confuse.

Model Standards

A potentially significant change for the construction sector is that BSI has given up its unilateral resistance to the International titles of IS0 9001, 9002 and 9003. The title 'Specification' has been abandoned and the International title of 'Model for Quality Assurance...' has been adopted. The difference was highlighted in SP84 because it appeared to inhibit 'tailoring' of the Model Standards to the particular requirements of the construction sector, despite the express provisions of the Standards which allow for tailoring. Since the Model Standards have their roots firmly in the manufacturing and process industry sectors, it should not be surprising that some elements are inappropriate to construction and that tailoring is needed.

Within the three Model Standards, the amendments are not substantial. Many are at a level that a parliamentary draftsman might enjoy, e.g.:

- The numbering of clauses in IS0 9002 and IS0 9003 has been revised to make the three Model Standards consistent: that is quite a helpful, if rather superficial, change.

- What was in Notes has, in some instances, become main text, and what was in main text has been put as Notes. The significance of this is somewhat elusive.

It is not a criticism that the changes are minimal (except that people may feel cheated at having to pay the exorbitant prices for the new editions), since the 1987 edition was so well drafted. There are, however, a few more substantial changes. Legal considerations have been given some acknowledgement. Para. 4.5.2(c) now recognises the possible need to retain superseded documents for legal and related purposes, and not to destroy all copies. It states:

> "The control shall ensure that...(c) any obsolete documents retained for legal and/or knowledge preservation purposes are suitably identified."

This meets one of the key recommendations in SP84 that "The potential use in legal proceedings of documents prepared for (or generated by) a quality management system should be appreciated". The amendment is, unfortunately, not reflected in the Guidance document, ISO 9004-1, Section 17.2, Quality Records. A second nod to legal implications is that there is now recognition, for the purposes of 'contract Report', that transactions commonly involve 'tenders' as well as 'contracts', and that there may be amendments to a contract. The definition of 'tender' is almost, but not quite, comprehensible. It reads:

> "Offer made by a supplier in response to an invitation to satisfy a contract award to provide product".

However, the reluctance of the quality management world to allow significance to the law is displayed in the definition of 'contract', which reads:

> "Agreed requirements between a supplier and customer transmitted by any means".

There is, of course, the difficulty that the Standard has to be applicable in all parts of the world, while law differs from place to place. The concept of a contract as a legally binding agreement is practically universal, but the conditions for an agreement to be legally binding are not. The definition works in some contexts, but not others, For example, para. 4.16 reads: "Where agreed contractually, quality records shall be made available for evaluation by the customer or the customer's representative for an agreed period". Such a usage only has meaning in terms of a legally-binding agreement.

The design stage is addressed only in the first of the Model Standards, ISO 9001. The relevant Section 4.4, Design Control, has been substantially revised and extended. In particular, the element of design verification has been replaced by three elements: design Report, design verification and design validation. The purpose of design validation is described as "to ensure that product conforms to defined user needs and/or requirements" and Note 13 states that "Validation is normally performed on the final product". The application of this element to design and production of a refrigerator or car is clear, but its application to construction works is not. It suggests full-scale testing of the completed works. More importantly, it implies at least ongoing involvement of the designer during the construction phase until completion. This is incompatible (perhaps rightly so) with the trend of some promoters to relegate designers to 'design contractors'.

The problem, identified in SP84, of the application of ISO 9001 section 4.4 to a design-only organisation has not been addressed. Para. 4.4.1 still requires "procedures to control and verify the design of the product". If this is applied to a design-only organisation and the 'product' is interpreted to mean the product of that organisation, the requirement makes nonsense. Design-only organisations will still finish up trying to apply quality system elements to design processes which were only intended to apply to production processes. The problems should not arise since the Model Standard is specifically directed to organisations which undertake all stages, from design, through production, to installation and even servicing. The 'product' is the result of all stages. Nevertheless, the insistence of UK certification bodies that ISO 9001 applies to consulting engineers and architects continues to create difficulties. (If certification bodies correctly operated quality systems in relation to their own activities and responded to criticisms, the problems should not still exist.)

Section 4.9, 'process control', has been revised, apparently to reduce the significance of the description 'special process'. The new edition merely notes that processes which require prequalification of process capability are frequently referred to as special processes, whereas the previous edition appeared to make the requirement of such prequalification dependent on whether the process was classified as a 'special process'. There is also an interesting new requirement in section 4.9(g) for "suitable maintenance of equipment to ensure continuing process capability".

Guidance Standards

The most extensive changes in the new edition are in ISO 9000-1, Guidelines for Selection and Use. It is difficult to say whether the changes are important since, in principle, the Guidelines are not part of the Model Standards and do not create auditable requirements. Nevertheless, the changes may give concern as signalling a shift towards excessive idealism and enthusiasm in place of experience and realism. Quality management converts have always exhibited evangelical tendencies, but there is the danger of persons in positions of effective power slipping into a cult of unquestioning fervour. ISO 9000-1 now talks of the "ISO 9000 family" and claims the Guidance document has the "role of road map for the ISO 9000 family".

The document has been expanded substantially from the 1987 version. Additional 'guidance concepts' have been included, said to be:

> "- *needed for effective understanding and current application of the ISO 9000 family; and*
> - *planned for complete integration into the architecture and content of future revisions of the ISO 9000 family.*"

An example of these concepts is in paragraph 4.1(b):

> "*An organisation should...(b) improve the quality of its own operations, so as to meet continually all customers' and other stakeholders' stated and implied needs.*"

When one reads in 4.2 that 'stakeholders' are envisaged to comprise the organisation's customers, its employees, its owners, its sub-suppliers and society, and that "The supplier should address the expectations and needs of all its stakeholders", one begins to understand why an employer, such as BSI itself, might find it uneconomical to employ an organisation burdened by a quality system to carry out a simple job such as painting. 'Quality' appears to be set to take on a whole new meaning.

A less cynical response would be to say that such wider concerns are becoming part of the accepted business environment. The recent judgment of Sir Godfray le Quesne QC (sitting as a deputy High Court judge) in *General Building & Maintenance plc v Greenwich Borough Council*[5] demonstrates that these wider concerns may properly be considered by public authorities, for example in applying the Public Works Contracts Regulations 1991 in deciding whether to admit a contractor onto a tender list on the grounds of technical capacity:

> "*Technical capacity, in my judgment, means ability competently to carry out the necessary operations of the contractor's trade. For many contractors such as the plaintiffs, ability competently to carry out the operations*

---

[5]   (1993) 65 BLR 57

of their trade in these days includes ability to carry them out with proper regard for the health and safety of those whom they employ and members of the public whom they affect."

ISO 9000-1 paragraph 4.2 elaborates on the "requirements of society" and its vision of the role of quality systems in regard to those requirements, as follows:

"The requirements of society, as one of the five stakeholders, are becoming more stringent world-wide. In addition, expectations and needs are becoming more explicit for considerations such as: workplace health and safety; protection of the environment (including conservation of energy and natural resources); and security. Recognising that the ISO 9000 family of International Standards provides a widely used approach for management systems that can meet requirements for quality, these management principles can be useful for other concerns of society. Compatibility of the management system approach in these several areas can enhance the effectiveness of an organisation."

The issue is one of boundaries. If quality systems do not maintain closely defined boundaries as to both their own ambit and the requirements which they seek to satisfy, they will lose their focus and costs will be incurred which reverse the financial effectiveness and advantage of implementing a quality system. The requirements of society are primarily a matter to be established by law-makers, rather than by exhortation in Standards. It may be appropriate for some organisations to anticipate, and even pre-empt, legislative trends but, for most organisations, to do so is likely to be commercial suicide.

Another concept developed in ISO 9000-1 is that of 'processes', particularly in sections 4.6, 4.7 and 4.8. This states:

> *"An organisation needs to identify, organise and manage its network of processes and interfaces. The organisation creates, improves and provides consistent quality through the network of processes. This is a fundamental conceptual basis for the ISO 9000 family"*

Changes to reflect this new conceptual basis appear, for example, in the section headings in ISO 9004-1: 'Quality in production' has become 'Quality of processes'; 'Control of production' has become 'Control of processes'. It remains to be seen whether the new conceptual basis proves effective, but the change emphasises the possible dichotomy between contractual obligations which define end-product requirements, and quality system elements which define process requirements. Unfortunately, the reference to management of interfaces is not elaborated. There is still no assistance on the management of interfaces between organisations which do not come within a unified quality system. This problem was identified in SP84 as fundamental in the construction sector when the conventional division of design and construction is employed.

Finally, ISO 9000-1 resolves one argument which took up much time in the Steering Group meetings for SP84. 'Quality' does not include price. Note 15 states unequivocally:

> *"Product value involves both quality and price and, as such, price is not a facet of quality."*

The second Guidance Document, ISO 9004-1 has not been amended to the same degree. There is a new division of 'configuration management', which may have some relevance. More substantial is the introduction of a section, 'Financial considerations of quality systems', which states that it is important that the effectiveness of a quality system be measured in financial terms. This may be regarded as an antidote to the wilder excesses of idealism in ISO 9000-1. Also significant is the revision of section 8, 'Quality in specification and design'. This ties in with the amended section on design control in ISO 9001.

Finally, one might pick out an additional sentence in paragraph 15.5, under the heading 'Analysis of a problem', which may be particularly relevant to investigations of construction quality problems, discussed below. This states:

> *"Consideration should be given to establishing a file listing nonconformities to help identify those problems having a common source, contrasted with those that are unique occurrences".*

Matters not taken on board

The big weakness of the ISO 9000 series for the construction sector is revealed in Note 1 in ISO 9000-1, which states:

> *"In all these International Standards, the grammatical format of the guidance or requirements text is addressed to the organisation in its role as a supplier of products."*

It was pointed out in SP84 that there was a need, in construction, for a Model Standard for Procurement, addressed to employer organisations. The 1994 revision of ISO 9000 has emphasised the unwillingness of the Standards bodies to accept that quality and risk control can depend to a significant degree on the role of the ultimate procuring organisation.

SP84 recommended strongly that there should be a separate Model Standard for design-only organisations, to avoid the problems that result from trying to apply ISO 9001. No progress appears to have been made on that, although it is a very simple matter. SP84 also recommended the drafting of standard quality system elements to address the problems of existing ground conditions and existing structures. These matters do not fall comfortably within the ISO 9000 approach, which is predicated on the basis that everything affecting quality is either the supplier's or the customer's product. This problem is one strictly for the construction sector to resolve.

Latent Defects Insurance and Construction Quality Forum

Although the impact has been concealed by the depressed state of the construction market for new building, considerable progress has been made over the past two or three years in the availability of latent defects insurance for buildings. Latent defects insurance was strongly advocated by SP84 as a natural companion to the development of quality management systems, to provide 'quality assurance' in all senses. SP84 pointed out that latent defects insurers would be attracted by the adoption of quality management systems in construction.

SP84 also advocated the development of a system to collect data on construction defects and problems, and anticipated this might fit in with the spread of latent defects insurance, as insurers became interested in the elimination of defects. Although this was not an original idea, it is particularly pleasing to see that, in November 1993, the Construction Quality Forum was set up under the aegis of the Building Research Establishment to receive and disseminate information on problems encountered in construction.

The Forum is supported by members who are expected to make reports on problems encountered, with a view to receiving the analysed information back in due course. It is early days yet to predict how the Forum will develop. Initial promotional literature still concentrates on secondary or peripheral matters, such as the use of an optical scanner to read the reports - one suspects that the degree of detail collected in the reports will need to be enhanced if worthwhile conclusions are to be drawn and recommendations made. Questions arise as to the potential imbalance in costs and benefits between those who have the most information, such as specialist expert witnesses, and those who will be able to use the information: designers, constructors and owners. Is there sufficient incentive for the key potential sources to provide the information? Also. who is to release them from duties of confidentiality?

The possible link with ISO 9004-1 para. 15.5, set out above, should be pursued. The construction sector will only overcome quality problems if there is co-operation between organisations to identify and solve them.

Quality Management and Risk

Coming at last to the title subject of the paper, it is reasonably well known that the underlying raison d'être of quality management systems is the control of risk in regard to defective product. Reductions in defect-related costs are expected to provide savings many times the cost of implementing a quality system. There are, however, other aspects of risk. SP84 warned of dangers when implementation of a quality management system is made a contractual requirement. It introduces new risks:

- the risk that stifling or excessive procedures will be imposed;
- the risks that a client's representative will not be capable of dealing rationally with concessions in respect of minor nonconformities;

- the risk that an undertaking to implement a quality system will be interpreted as raising a skill and care obligation to one of fitness for purpose.

The last-mentioned risk is considered to be an illusion, but avoidance of the first two depends on attitudes and understanding. Just as it is possible to apply any form of contract to produce either a happy or an unhappy result, the same is true of quality systems. The construction sector needs to be continually alert to prevent negative outcomes. In particular, there is a need to establish clear boundaries for quality management systems, both as to their scope and their objectives

What is most needed, however, is the application of quality systems at the pre-construction contract phase, to minimise the risks, rather than share or transfer them. Quality system requirements could play a significant role in ensuring adequate and appropriate pre-contract ground investigations or investigations of existing structure. They could be applied to reduce the extent of ambiguities and discrepancies in contract documentation. Those responsible for Quality management Standards must be persuaded to accepted the validity of Sir Michael Latham's statement: Implementation begins with clients. Quality management should also begin with clients.

7. The Competitive Procurement of Professional Services

John Connaughton

Synopsis

This paper, based on the results of a research project undertaken for CIRIA[1] between 1992 and 1993, summarises the main recommendations of the recently published CIRIA guide[2] to the use of competition for the procurement of construction professional services. A historical overview is presented, followed by detailed recommendations based on a distillation of current good practice. It is concluded that no single prescription exists for the competitive selection of consultants.

Historical background

Two reports from the Monopolies and Mergers Commission[3] in the 1970s challenged the role of the professional institutions in

---

1   Davis Langdon Consultancy, *Obtaining Quality and Value Through Competition in the Procurement of Professional Services*, Final Report of CIRIA research project RP 464, CIRIA, London, 1993.

2   Connaughton, J.N., *Value by Competition: A Guide to the Competitive Procurement of Consultancy Services for Construction*, SP 117, CIRIA, London, 1994.

3   Monopolies and Mergers Commission, *A report on the general effect on the public interest of certain restrictive practices so far as they prevail in relation to the supply of professional services*, Cmnd 4463, HMSO, London, 1970; *Architects'/Surveyors' Services: a report on the*

the recommendation/prescription of fee scales as the charging basis for professional services. During the late 1970s and early 1980s, the institutions increasingly found themselves on the defensive against the charge of operating restrictive (and monopolistic) practices for the benefit of their members.

Then, as now, the institutions' position was weakened further by an economic recession which contributed market pressure to the growing trend away from the use of prescribed fee scales. The Monopolies and Mergers Commission - despite strong objections from the professional institutions - concluded that the imposition of any restriction on competition on the basis of fees was against the public interest. By 1981 the then Conservative Minister for Consumer Affairs gave notice to the professional institutions that, unless they gave certain undertakings with respect to fee scales, she would make an order under the Fair Trading Act 1973.

By early 1982 the institutions had begun to take action, involving in many cases the abrogation of mandatory fee scales and the abolition of bye-laws and other regulations which, for example, prevented members from quoting fees in competition. Other 'restrictions' deemed not to be in the public interest were revised or withdrawn. (The RICS continued to publish fee scales for quantity surveying services, for example, on the basis that these were 'recommended' and not mandatory). Mansfield et al[4] argue that it was the recession in the early 1980s which accelerated the onset of open competition within the professions for the provision of consultancy services. The greatest changes in the procurement of professional services have occurred in the public sector. By 1984 the Property Services Agency (PSA) had, with its Accommodation and Works Circular 22/84,

---

supply of architects'/surveyors' services with reference to fee scales, HMSO, London, 1977.

[4] Mansfield, N.R., Rowdon, I.J. & Dunn, M.K., *The unfolding saga of fee competition*, Civil Engineering, June 1988, pp.34-5.

introduced fee competition for all consultant commissions (except term commissions) with the intention of applying fee bidding to all projects above £0.5m. A review of this practice was established in 1985 and completed in 1986.[5] The review found that competition had led to substantial reductions in fee levels for all types of commissions. and that in general the level of claims or instances of poor quality had not increased significantly as a result. The report concluded that the Agency should continue with this form of procurement.

Other government departments claimed similar experience. By the mid 1980s almost all central government departments were procuring professional services through competition. Additionally, many of the public utilities and statutory undertakings, for example, British Telecom, British Gas and the water authorities, adopted competitive procedures for consultant appointment and continued to apply these following their privatisation during the latter part of the 1980s and early 1990s. The use of competition by these bodies increased as they divested themselves of a large component of their in-house professional resources and came to rely more and more on external consultants.

Recent developments

Towards the late 1980s the procurement of professional services through competition became more widespread in the UK public sector. The deepening recession of the late 1980s and early 1990s again led to a keenly competitive market generally for consultancy services. Whilst it had generally been the policy of government to consult the professions about changes in government's procurement practices - the introduction by the

---

[5]   Property Services Agency, *PSA Fee Competition Review Committee Report*, HMSO, London, 1986.

PSA of fee competition in 1984 had been preceded by extensive negotiations with the major professional institutions on fee levels, conditions of engagement and selection procedures[6] - government nevertheless maintained that it was up to the relevant departments to act in a manner which best protected their interests.

Most Government departments which had introduced fee competition in the middle of the decade had, by the late 1980s, lowered the contract value thresholds above which competition was applied. Furthermore, many procurers, particularly in local government, sought to exploit a 'buyer's market' for consultancy services by inviting large numbers of bidders to compete for all but the very smallest commissions on the basis of lump sum fee bids. Although practice varied widely, there was a significant departure from earlier PSA recommendations, for example, that a maximum of three suitable consultants be invited to fee-bid.

Consultants and their representative professional institutions were becoming increasingly critical of the use of competition, particularly when selection was based solely on price. They were concerned with the erosion of remuneration for professional services and they also claimed that this would undermine professional standards to the long-term detriment of their clients. The institutions produced their own forms of guidance for clients on competitive procurement generally and on fee competition in particular[7], arguing that whilst they were not opposed to competition per se, they were opposed to selection

6   Barnes, A., *The Battle Over Fees and Competition*, RIBA Journal, May 1985, pp.76-80.

7   See, for example, Association of Consulting Engineers, *Advice to Clients on Fee Competition*, ACE, London, 1992; Royal Incorporation of Architects in Scotland, *Code of Procedure for Fee Tendering*, RIAS, Edinburgh, 1987; Royal Institution of Chartered Surveyors, *A Guide to Fee Tendering for Building Surveying Services*, RICS, London, 1985.

on the basis of price alone. Guidance (from the professionals' perspective) was provided for local authority procurers by the Society of Chief Architects in Local Authorities (SCALA)[8] in the form of a recommended code of procedure. This underlined earlier PSA guidance in recommending that fee competition should generally be held with a maximum of three suitably qualified consultants.

In an attempt to consolidate guidance on good practice in the public sector in consultant selection and appointment, HM Treasury's Central Unit on Purchasing (CUP) published its Guidance Note 13[9] on the appointment of works consultants in 1989. This claimed to provide a statement of good professional practice in consultant appointment to "all types and sizes of projects". The Department of Transport, which introduced fee competition for consultant appointments in 1985, also reacted to growing criticism of its fee competition procedures around this time. The Department revised its procedures in 1990 is response to the need for clearer briefing and the introduction of technical merit in the evaluation of consultants' tenders.

However, the government pursued its policy of introducing competition to all aspects of public services and published a White Paper "Competing For Quality"[10] in 1991. This proposed extending Compulsory Competitive Tendering (CCT) to professional and technical services. It attracted considerable opposition from the professional institutions, some of whom saw no let-up in the Government's attempts to introduce fee competition into all areas of professional practice Although

---

8   Society of Chief Architects in Local Authorities, *Recommended Cods of Procedure for the Commissioning of Professional Services for Building and Related Works,* Occassional Paper No 47, SCALA, 1985.
9   Central Unit on Purchasing, *CUP Guidance Note 13: The Selection and Appointment of Works Consultants,* HM Treasury, London, 1989.
10   Department of the Environment, *Competing for Quality: Competition in the Provision of Local Services,* HMSO, London, 1991.

guidance from the Department of the Environment[11] reaffirmed that government policy on consultant selection was as outlined in CUP Note 13 - and was concerned primarily with quality and value for money in the design of government buildings - the professional bodies continued to be concerned about the widespread use of fee competition.

The deepening recession of the early 1990s again led to reductions in the remuneration levels of consultants and, indeed, threatened the very survival of many. The professional institutions, acting through the Construction Industry Council (CIC), supported draft guidelines for the procurement of professional services through competition.[12] The CIC proposals, with reference to the USA Brooks Act of 1972[13], endorsed a 'two-envelope' bidding system, and have subsequently been revised to include a proposed marking system for the evaluation of consultant's technical proposals.[14] Other recent proposals published by the professional bodies or with their support, advocate the use of a systematic method of evaluating consultants' technical proposals and/or abilities as an essential part of the competitive selection process.[15]

---

[11]   Department of the Environment, *Guidelines for the Design of Government Buildings*, DOE/HMSO, London, 1991.

[12]   Construction Industry Council, *The Procurement of Professional Services: Guidelines for the Application of Competitive Tendering*, CIC, London, 1992.

[13]   For a description of the Brooks Act and its operation, see Lorenze, B.A., *The Brooks Architect/Engineer Act: Promoting Quality in Federal Contracting*, American Consulting Engineer's Council, Washington DC, 1993.

[14]   Construction Industry Council, *The Procurement of Professional Services: Guidelines for the Value Assessment of Competitive Tenders*, CIC, London, 1993.

[15]   See, for example, Architecture Unit of the Arts Council of Great Britain, *Architecture and Executive Agencies*, Arts Council of Great Britain, London, 1993; Association of Consulting Engineers, *Balancing Quality and Price: Value Assessment and the Selection of Consulting Engineers*, ACE, London, 1993; and Royal Institute of

Competitive procurement: current good practice

The majority of respondents to a survey undertaken on the CIRIA study[16] felt that the use of competition was set to increase in the future. None of the public sector informants questioned during the research foresaw that their procurement policies would move away from the use of competition. Indeed, many considered that, as they became more used to specifying the services required, so their need to consider consultants' abilities at the final selection stage diminished. In the main, private construction clients have not adopted competition for the procurement of professional services as rigorously nor as extensively as their public sector counterparts. However, there is a growing need in the private sector for selection procedures to be open and accountable, and many private sector procurers appear to be moving towards some form of competitive selection.

The author believes that the CIRIA study and forthcoming guidance are being made available at a time of transition regarding policy and attitudes (both public and private) towards competitive procurement. However, current procedures used have not yet matured and gained the status of well established, 'generic' procurement methods, as is the case with construction work generally. Indeed, current practice is evolving as both clients and consultants become used to operating what was, until relatively recently, an unknown form of commissioning construction services. What follows is a distillation of the more workable and equitable elements of current practice. However, it must be noted that, given the widespread and at times acrimonious debate about the use of competition for the procurement of professional services, there is little consensus

---

British Architects, *Recommended procedure for the selection of a chartered architect*, RIBA, London, in preparation.

[16] Davis Langdon Consultancy, *op. cit.*

between clients and consultants about what constitutes good practice.

In the absence of any well established ground rules, it is useful to note that there is broad agreement about a number of aspects of competitive selection. These are covered more fully in the study final report.[17] In particular, there is:

- Little opposition to the principle of competition
- a general lack of support for open competition
- widespread use of two-stage tendering
- recognition of the importance of consultant's ability as a selection criterion
- recognition of the importance for many clients of price.

Quality by Competition

What follows is a précis of the key elements of the CIRIA guide.[18] It is beyond the scope of this paper to list all the guide's recommendations. While the main points are outlined, the coverage is necessarily selective. The reader is referred to the full guide for further information.

Deciding to use competition

There are six key steps in the competitive selection of consultants:

- Decision;
- Preparation;
- Pre-selection;

---

[17]   Davis Langdon Consultancy, *op. cit.*
[18]   Connaughton, J.N., *op. cit.*

- Specification;
- Selection; and
- Award and Appointment.

The procurer's aim is to find the most appropriate consultant in terms both of ability and, when important, price. Competition has three significant advantages over negotiation in this respect:

- First, competition encourages a systematic approach which is more likely to yield the right person than negotiation with a single consultant;
- secondly, competition is a transparent process and so helps satisfy the increasing demand for accountability in both the public and the private sectors; and
- thirdly, competition can give clients better value for money, particularly when consultants are compared on the basis of both their ability *and* their fees.

Very few of the clients, consultants and contractors consulted during the research were opposed to the principle of competition. However, the use of consultants' fees as a selection criterion remains hotly disputed. The arguments put forward against competition therefore tend to focus on this aspect. In summary, it is claimed that:

- First, the value of a consultant's contribution is more truly judged by his ability than his fee. Fee competition forces consultants to do the minimum necessary with the result that the client does not get the most valuable services - nor the best project.
- Secondly, the professional services needed at an early stage in the construction process cannot be adequately described to allow for fair competition. This is because they include intangibles, such as creativity and judgement, and because they help define the works ultimately needed. Until the scope of

the works is clear, that of the services required to define them will be uncertain.
- Thirdly, competition is a time-consuming and an expensive process and its costs may well outweigh any fees saved.

In refuting the case against competition, the following are emphasised:

- First, ability is the prime criterion when selecting the right consultant, but price is also important for many clients. Only by taking account of ability and price can procurers be sure that the most suitable consultant is selected. Capable procurers will recognise the importance of weighing the potential value of consultancy services against their price, and will not settle for a substandard service just to save a small amount on the fee.
- Secondly, it is difficult to define precisely what consultancy services are needed, whether or not competition is used. Competition requires a clear and comprehensive specification of services.
- Thirdly, competition takes time and costs money but the benefits of finding the most suitable consultant for the job far outweigh this initial investment.

The practical exceptions to the use of competitive tendering are where the likely benefits from competition would not outweigh the costs, or only one consultant has the specialist expertise needed. Similarly where services are needed urgently and there is not enough time to undertake the competitive process properly this method of tendering should be avoided.

The form of competition

Competition may be either open or selective. Open competition is sometimes preferred over selective competition because it offers a wider choice and is generally less open to charges of favouritism. It may also offer consultants an otherwise unavailable opportunity of bidding for a particular appointment. But the potentially large number of candidates makes open competition expensive for both clients and consultants if the bids are to be properly prepared and evaluated. For this reason, selective competition is generally preferred. A two-stage approach is recommended. The first stage involves the pre-selection of consultants to form a short list from whom final bids will be invited. The objective is to ensure that all short-listed consultants are competent to carry out the work. In summary final selection should be preceded by pre-selection and both stages should be restricted

The basis of competition

The bases on which consultants compete are on the consultant's ability only, or the price for the consultancy services, or maybe a combination of both. The main advantage of ability as a selection criterion is that it helps ensure that the best qualified consultant is chosen. But it ignores the price of the consultancy services - which can vary. Competition on ability alone will not necessarily lead to the appointment of the most suitable consultant in terms of both ability and price.

Competition on ability and price is where selection is based on a combined assessment of the consultant's ability and fee proposals. Competition on ability and price offers the potential for rigorous selection procedures which are open and accountable and help to demonstrate value for money, and most importantly help in selecting the most suitable consultant in terms of both ability and price.For this basis to work, procurers must genuinely be able to trade-off the consultant's ability against the price of his service.

Competition on price is apparently the simplest basis on which to select consultants, and the most contentious, since, if it is used as the final criterion, procurers must be confident that price alone differentiates the competitors. The obvious advantage of price as the final selection criterion is that it obtains the lowest price for the required services. It also helps avoid allegations of favouritism. The main disadvantage is that, if not preceded by pre-selection on the basis of ability, the value for money of the consultancy services cannot be assessed. Price alone should only be used at the final selection stage when the services needed can be clearly and comprehensively defined and there is otherwise little to choose between competing consultants.

Even at the final selection stage, there are problems with price as the sole criterion. First, to enable all consultants to compete on price there is a risk that procurers will settle on a minimum acceptable standard of service as the benchmark. This may preclude the offer of more valuable - though more costly - services which may be of greater benefit in the long run. Secondly, consultants' price offers may vary considerably and, in a depressed market, some may bid below cost just to continue working. They may also attempt to 'buy' a commission for a low bid in the hope of securing more work from the same client at a higher price. Consultants who work below cost may only be able to do the bare minimum; there is also a risk that they may try to recover costs by seeking opportunities to claim for extra work.

Preparing for competition

Competition means that procurers must understand client requirements in order to identify and pre-select the most suitable consultants and document these requirements in a form which consultants can understand and compete on. Procurers are entrusted to organise and manage the competing proposals and on the basis of the clients requirements to make an award. Proper planing and preparation is essential to good procurement practice, whether or not competition is used. While competition requires clients and their procurers to think about their needs rather earlier than is the case with negotiated appointments, this can be a helpful discipline and client requirements can be clearer as a result. Guidelines on the specification of consultancy services are beyond the scope of this paper but are covered in the forthcoming CIRIA guide.

The importance of pre-selection

Experienced procurers recognise that the more care is spent on pre-selection, the greater likelihood there is that the consultant finally selected will be the 'right' one for the job. Pre-selection ensures that suitably qualified consultants are identified for final competition and that the competition is kept to manageable proportions. Both clients and consultants incur costs in competition; the more competitors there are, the higher these costs will be. The key steps in the pre-selection process are:

- Identify requirements;
- Compile the initial list of consultants;
- Compile the long list of consultants;
- Confirm the interest of 4-6 consultants;
- Identify the pre-selection criteria;
- Invite pre-selection statements;
- Evaluate pre-selection statements;
- Select 3-4 consultants for final competition; and
- Inform all consultants of the outcome of the pre-selection.

Criteria for pre-selection depend largely on how clearly clients' requirements can be defined. The clearer these are, the more focused, rigorous and comprehensive the preselection evaluation can be. In general, pre-selection should focus on the consultant's relevant experience and qualifications. However, if client requirements are already clear at the pre-selection stage, consultants may be asked for suggestions as to how they would approach the work.

The key criteria for pre-selection are as follows:

- track record on similar commissions, in terms of:
  - approach;
  - technical ability;
  - performance to quality, time and cost constraints;
- professional resources and support facilities available;
- qualifications of key staff;
- financial standing;
- adoption of a quality management system; and, if appropriate
- outline suggestions for approaching the commission, in terms of:
  - experience of proposed the project manager/team leader;
  - how the commission will be organised and managed;
  - technical, managerial or design method.

There are a number of ways in which procurers can obtain information from consultants to help evaluate their ability at the pre-selection stage. A general invitation as to the consultants statement of interest is the preliminary stage in assessment. If the consultant is of preliminary interest examples of the consultants' previous work could be viewed and their offices visited. An interview with the consultant and taking up of previous client references may be of great use. The assessment of a consultants ability necessarily involves making value judgements. At this stage there is unlikely to be enough information for a formal evaluation. It is therefore wise to concentrate on seeking answers to the following key questions:

- Is the consultant capable of providing the services needed?
- Is his track record satisfactory?
- Are the necessary staff and other resources available?
- Are any particular requirements satisfactory?
  and, if appropriate:
- Is his proposed approach to the commission acceptable?

Final selection

Tendering procedures must be, and must be seen to be, fair and workable. It is important that the consultants have confidence in the selection process, that the selection process produces tenders that are useful and comparable and that ultimately the most suitable consultant is selected. The essential principle is to treat all tenderers equally. The documentation must be common to all tenderers and in cases where, for example, an individual consultant queries about the works or services required the response to this information request must be sent to all the tenderers. The timetable for issue, receipt, adjudication and award of tenders must also be strictly adhered to.

Requesting consultants to submit their proposals under specific headings, as suggested, indicates broadly which criteria will be used for the evaluation, as a result the tenders will be more focused and directly relevant and will be easier to compare.. Some procurers claim that this is enough, arguing that consultants should concentrate on the requirements of the project rather than tender evaluation. Consultants argue that if they knew how their tenders were to be evaluated they could supply more directly relevant information and the tendering process would be fairer. The difficulty is that consultant selection inevitably involves a degree of subjective judgement about the relative importance both of different selection criteria and of ability and price. But while procurers must have some freedom to vary these criteria from project to project, there are advantages in informing consultants how their tenders are to be evaluated.

It is recommended that tender documentation should make clear to tenderers, the basis of competition, the criteria to be used in the evaluation of ability and the priority order to be given to these criteria. In the case of competition on ability and price, where ability is to be scored and used to weight the fee bid, an indication should be provided of the minimum weighting for ability. It is not recommended that either the proposed marks for each criterion, or the precise weighting for ability be divulged. To do so would limit the procurer's freedom to choose, for example, a proposal of outstanding quality on one or a number of key criteria, or an extremely good but expensive proposal, whose worth to the client far outweighed its cost.

When the interview stage is reached, it is essential that the consultant is represented by people, especially the team leader, who will undertake the commission rather than solely by senior staff who will have little direct involvement in the work. Interviews are not normally necessary where the project is simple and straightforward, or the services are clearly defined. Similarly where the client or procurer knows the consultant well interviews may not be necessary

Effective pre-selection should mean that all the consultants who have got this far are capable of undertaking the commission. An assessment of their overall experience and available facilities need not dominate the final selection.

Evaluating fee bids: Fee levels and resources

Fee levels at any given time will be influenced by a variety of factors, including:

- the market for the consultancy services required
- the consultant's assessment of the scope and level of services he must provide, and the risk he must bear
- the consultant's current and likely future workload
- the consultant's *modus operandi,* and whether, for example, a 'ready-made' team is available
- the consultant's experience of working with the client and other project participants.

It is worth bearing in mind that while procurers may be attracted by a low fee bid, one which is 'too low', where the consultant will be working at a loss, can lead at best only to the necessary minimum of service and at worst to substandard performance and unwarranted claims for additional payment. However, it is difficult to determine when a bid is too low. Consultants' costs, profit levels and attitudes to risk vary considerably, as do their opportunities for economy in the services they provide. Where the consultant ranked highest on ability submits the lowest fee bid, then subject to the proviso on fee levels above, he is the first choice. But usually procurers will have to combine each consultant's fee bids with the assessment of his ability (or vice versa) to identify the best value for money offer.

Procurers should request a statement of the staffing levels and other resources which consultants are to provide for the fee quoted. This will help establish whether consultants are proposing comparable levels of resources and that their average time and 'real' time charge rates, i.e. the total fee divided by the staff resources proposed, are comparable to quoted rates for time-charge or additional work, are similar

There are two main approaches to combining ability and price. The first is a simple weighting which uses the aggregate ability score, out of, say, 100, to weight the price bid. The second involves allocating prior weightings to ability and price. Aggregate scores for both ability and price are then calculated and multiplied by their respective weights to determine an overall score. There is no perfect or 'correct' method of combining the evaluation of ability with consultants' fee bids. Wide fluctuations in fee bids, for example (assessments of ability are more likely to fall within a narrower range), may upset any pre-determined weightings for ability and price. Arithmetic methods are only really useful as an aid to the selection of the most suitable consultant. The final choice will involve considerable value judgements by client and procurer.

Making the award

At the interview, clients and procurers should have satisfied themselves that the preferred consultant has a good understanding of all aspects of the commission, the necessary capability, staff and resources available for the duration of the product. It is good practice to choose the consultant on the day that interviews are held. Details of the evaluation will still be fresh and the selection panel will be present to debate and finalise particular preferences and choices, weighing potential benefits of the award against the fee.

Once the consultant has confirmed that he will accept the commission under the terms set out in the tender documentation, other tenderers can then be told that they have been unsuccessful, indicating the name successful tenderer and briefly why that tenderer has been successful. Mention should also be made of the range of fee bids received, and the names of the unsuccessful tenderer. This is a frequently overlooked aspect of tendering practice but is essential if consultants are to have confidence in the process.

Conclusion

The foregoing description of competitive selection offers a brief look at some of the key recommendations of the CIRIA guide. It is not possible here to re-iterate all of the guide's recommendations, but three core principles may be noted in conclusion:

- First, value for money is not to be confused with cheapness. The aim of the procedures outlined here is to select and appoint the most suitable consultant. This may mean paying more than the minimum for valuable services. But it does not ignore that many clients are constrained in how much they can afford.
- Secondly, time and effort invested early in the competitive process - in particular in specifying the services needed and pre-selecting consultants - will be amply repaid later; and
- Thirdly, consultant selection requires that clients make value judgements about the worth to them of the services offered for the fees quoted.

It is hoped that the CIRIA guide will help in these decisions, but ultimately clients must decide for themselves which services offer best value for money. There is no completely mechanistic means of consultant selection.

8. Procurement of Construction Work The Client's Role

Malcolm Potter

Synopsis

The role of the client in the procurement process is considered first, as the individual patron of his own project and, secondly, as the catalyst for change and influence upon the industry as a whole. The responsibilities of the client are examined in the context of the different choices of procurement path. The influence of the client for good or for ill is illustrated through examples which show how individual and corporate actions and reactions bear upon performance of the industry. The paper raises some key issues which require colaboration between industry and client.

Introduction

Patron or patronised?

Ask any construction professional who is meant by 'The Client' and you will probably get some sort of answer like 'He's the one who pays'. But is that all he is? A recent survey canvassing views within the industry about the role of the client in the procurement process highlighted remarkably divergent opinions among the construction experts. Whilst some welcomed the client taking an active role, particularly if technically proficient, many felt that 'the client should not and need not get involved with the details of the project and should appoint competent consultants and stand aside from the process'.[1]

These conflicting views probably reflect reasonably accurately the differing attitudes about the client's role that exists within the United Kingdom industry today. Undoubtedly these differences can be explained, in part at least, by the great range of attitudes and competence displayed by the industry's clients. At some time or another most people in the construction business have experienced the confusion caused by clients who cannot stick to their decisions or who try to short-cut the agreed communications network. The same people may be less keen to admit that some resistance to the pro-active client may stem from the defence of traditional professional roles and demarcations which are in danger of being undermined or at least challenged by an outsider with the authority to 'interfere'. *Constructing the Team*[2] identifies the client as the 'driving force'. The question is, how can the client best exert this force: by 'standing aside', or getting involved?

The Client as the Patron for His Own Project

Who is the client?

Clients come in all shapes and sizes, from individuals to huge corporations or government departments. All except the very smallest make a functional separation between the decision making 'client authority', who may be represented by the Chief Executive, Board or Committee, and those who carry out the executive client function as Client Representative, Project

---

[1] CIRIA/ SBU Securing the contractor's contribution buildability in design (Draft) 1994

[2] Latham, Sir Michael, *Constructing The Team, Final Report of the Government / Industry Review of Procurement and Contractual Arrangements in the Construction Industry*, HMSO, London, 1994.

Manager, or Development Director. The term 'client' is used here to describe the 'client authority', *i.e.* the ultimate authority for providing funding, setting priorities, making decisions, and bearing the risks.

Many of those who find themselves in the role of client are not experienced or knowledgeable about the process of building procurement; their primary interests and expertise lie in the processes or tasks that require the building, and not the means of achieving it. Their motivation and imperatives may, therefore, at times be, quite understandably, at loggerheads with a sensible or appropriate procurement path, even if only to the extent that time scales or budgets may be unrealistically limited. Nevertheless, in the end it is this client who needs to understand sufficient of the building process and its risks to make the decisions and provide the resources necessary to ensure the satisfactory outcome of the project.

Clients' needs and priorities

How often in practice are the client's requirements fully met? The response to this has to be that the client's ambitions are very frequently not achieved, at least not in every respect. This can be just as true for the relatively experienced client as for the newcomer. Dissatisfaction arises for a whole variety of reasons ranging from minor time over-runs to major cost increases or even an unworkable building. How much responsibility, if any, should the client take for this?

It is common for consumers of the construction industry, and particularly dissatisfied ones, to compare the industry's performance with that of other industries (viz. Constructing the Team). Why, it is argued, can buildings not be produced on time, for a defined cost, well styled and engineered and backed by warranty, like a motor car? This is a 'chalk and cheese' argument. The comparison is unreal, simply because a new building is almost invariably a one-off prototype, whereas a motor car is the highly developed outcome of mass production. Only when the building prototype begins to emerge as a production model does the comparison have some validity. However, the analogy does serve to illustrate how the client, in this case the customer, is confronted by a similar set of considerations in selecting and purchasing a car, as is the prospective developer embarking on a new building project.

Ford's latest model is the product not only of continuous technical development and refinement but of considerable market research and feedback that takes account of the users' needs and aspirations. Modifications, for example, to include new and expensive safety features are introduced because people demand them and are prepared to pay for them. The customer also accepts that the design and production cycle takes time, often many years, and has to be prepared to wait for the development of a new features. He also expects the production to be carried out under controlled conditions by a trained and integrated team who have the expertise to produce a well designed and reliable machine. The price clearly reflects the range of choice and quality.

The customer makes his choice by careful consideration of his own needs and limitations, matching them to what is available. To do this, he takes account of delivery time, price (including methods of payment), running costs, appearance and performance (practical suitability, comfort, reliability, durability, safety and flexibility).

An informed choice requires active investigation and information about the products available. Most people would not dream of making an important purchase without considerable research. They would think nothing of spending a couple of hours going through the magazines before spending a few hundred pounds on a new television or stereo. How much more important this investment of time and effort must be when embarking on the procurement a new building costing thousands of pounds; a prototype, where the end product is uncertain, the means unfamiliar, and the production process is subject to a hostile physical and economic environment. It is a daunting task and unlikely to succeed unless the client, the motivating force, has at least a broad understanding of the responsibilities and tasks involved.

The extent to which the client needs to be involved in the building process depends partly upon his reasons for wanting to build; usually any one or more of the following:

- more space
- repair or improvement
- better fit
- improve accessibility
- develop/expand a business opportunity
- create a 'landmark'

And, like the car purchaser, he has his own particular priorities:

- Timescale - delivery time- critical/ non critical, short/ long-term
- Funding - price (including methods of payment), running costs, availability, timing, limits
- Quality - appearance, performance (practical suitability, comfort, reliability, durability, safety and flexibility)

These key factors shape both the building itself and the most appropriate procurement path .

The procurement paths - client's involvement and choice

Simply expressed, the client has two main choices of procurement path. He may either elect to deal with a single organisation (singlepoint) or he may appoint a number of separate ones, each responsible for an element of the process (multipoint). The singlepoint organisation takes responsibility for the complete design and construction operation and includes anything from simple package deal to complex design and build. The multipoint approach, which encompasses traditional contracting in one of its many forms, management contracting and construction management, requires the client to enter into a series of direct contracts with all those who are to take part. In his own interests he must ensure that these arrangements have interlocking agreements clearly defining the role of each participant. By implication this places the responsibility on the client for ensuring that the work of all these different people is co-ordinated. In practice, of course, much of this responsibility can be delegated. For example, in the case of traditional contracting the level of delegation to the design team leader is significant, whilst at the other end of the scale, under a construction management arrangement, the client himself may assume the role of chief co-ordinator.

Where the client's chief priorities are cost certainty combined with reasonable speed, he may opt for the singlepoint approach as the best means of securing his objectives. The use of singlepoint can have the advantage of increasing the cost certainty, but it does not eliminate the clients involvement in the process. Unlike traditional contracting where the architect or engineer assumes considerable management responsibility on the client's behalf, the singlepoint approach requires the client to provide an accurate brief, which may not be subject to later changes, evaluate its translation into the actual design, and deal directly with decisions he may have otherwise delegated to others. The singlepoint client's involvement in the procurement process therefore lies somewhere midway between that required for traditional contracting and the much greater commitment essential for the management type of contracts.

The singlepoint arrangement is unlikely to be the client's first choice where he wishes to exert a firm control over design. The multipoint arrangements offer the client greater flexibility in managing this part of the process and, for a premium, allow some degree of opportunity for changes at later stages of the project. In some cases a particular choice of procurement route will recommend itself as the best if not only approach to meeting his priorities. In others, several methods might be appropriate and the client makes his decision based upon a preferred or regular way of working that suits his organisation and available resources.

The effect of size/ experience upon the choice of procurement path and the client's role

The size, experience and expertise of the individual client has a profound effect upon both the choice of procurement path and the extent that can be involved in the process. For the purpose of placing this in a general context, clients can considered to fall into one of three types:

- an individual or organisation with a 'one-off' building requirement;
- organisations that regularly undertake developmen but rely wholly on external resources or agencies to implement the process; and
- organisations with major development programmes and directly employing technically skilled in-house resources able to both administer the procurement process and implement all or part of it as they choose.

Client 'A'

As a 'one-off' developer, client 'A' normally has little need or interest in acquiring development skills beyond those essential to manage his own project. Although most commonly responsible for smaller projects, this sector represents a relatively large proportion of the construction industry's workload. Individually being infrequent users of the industry's services, each new client is obliged to make use of the procedures, standards and practices as they are presented to him by his advisers. His lack of expertise means that he is less likely to want to become involved in the 'high risk' methods of procurement. The majority of clients in this category tend to opt for the more common traditional or straight-forward singlepoint arrangements.

Client 'B'

Client 'B' is able to take advantage of his more regular involvement with the building industry to develop informed and effective client skills. Even where the building programme is regular and quite extensive, many clients positively choose to restrict the use of dedicated in-house technical expertise to the client tasks of developing briefs and internal working practices; sufficient only to support the organisation's client function. This approach may favour a singlepoint approach to procurement particularly if where the building development has a functional continuity.

Client 'C'

Client 'C', is a major and regular developer who chooses to employ technical experts to implement his own projects. This is usually a conscious decision on the part of the client who has a particular reason for wanting control all of the process. Such organisations are capable of commissioning and controlling projects with considerable technical skill, using any of the industry's standard procurement systems and in some cases developing innovative technology and new ways of working.

Client tasks and duties

The role of client imposes an obligation to take responsibility for range of core tasks.

Client's primary tasks

These are the tasks that only the client can do himself. These cannot reasonably be delegated or shared with anyone else. Most also imply a duty to perform them responsibly. For example, there is little point in insisting upon priorities or targets that are counter-productive or unachievable by even the most positive or well considered strategy. Similarly it is sensible for the client to use his position of authority to promote co-operation and encourage the best performance from all those taking part. The scope of primary tasks include:

- deciding and explaining objectives, setting the achievable critical priorities;
- initiating feasibility and strategy review, understanding and accepting the implications and risks of the proposed procurement strategy;
- promoting an environment which insists upon good communications, clear lines of authority and encourages an integrated approach to the whole procurement (design and construction) process;
- setting criteria for selection and appointment of advisers and contractors;
- resourcing, through the provision of funding, information and other essentials including enabling through supportive authority;

- entering into contracts and accepting the responsibilities imposed by them; and
- reviewing performance against targets and taking action if not met.

Client's supporting tasks

These are the tasks which may in part be shared or delegated and with which in-house or external resources can provide assistance. Much of the day to day work involving detailed processing of information and decision making is often better undertaken by those with the time and expertise, provided, of course, that the work is co-ordinated. The scope of supporting tasks include:

- setting up internal arrangements necessary to investigate the feasibility and strategy of the proposed project;
- managing own internal organisation including information, communications and user advice and negotiation;
- selecting and appointing key advisers/ contributors;
- Providing briefing information (including advice not supplied by the building team);
- responding to routine questions and making day to day decisions within the client's established framework, including 'signing off' each stage as the project progresses;
- making payments; and
- monitoring targets and performance ;

The extent to which some or part of these tasks can be delegated to others varies with the task, the type of project, the chosen procurement route and the skills available to each client

a)   Deciding and explaining objectives, setting the achievable critical priorities

The client's objectives spring directly from his need to build. They may be modified by the feasibility process, but are unlikely to be changed unless the scheme is abandoned. They are the bedrock of the project and need to be fully understood by all those taking part. He must make sure that this is done.

The critical priorities which the client sets for his project determine the selection of procurement path. These may have to be modified by that choice once it is made. In practice it is often difficult, sometimes impossible, to find a procurement route that will meet all the preferred criteria. For example, high quality buildings cannot be built very quickly and cheaply: something has to give. The client has to be aware and decide precisely where his priorities lie and be realistic about their achievability: no-one else can do this for him.

b)   Initiating the feasibility and strategy review; understanding and accepting the implications and risks of the proposed procurement strategy

The client needs to be assured that the strategy for his project is based upon sound information and unbiased advice. The development of an informed strategy is a process that is too frequently 'glossed over' or overlooked altogether. Even for the experienced client it can be a difficult area, for the inexperienced client it can be a minefield. It is vital that the advice sought at this stage should be impartial and unaffected by the adviser's own self interest or limitations in experience, a view very strongly advocated by Constructing the Team. The client needs to know whether his project is viable and how best it should be approached.

There is no single or right way to obtain this information but in all cases the prudent client should be prepared to invest time, money and energy in exploring the options open to him. He must drive and control this process until he is satisfied that he has sufficient information and advice to enable him to decide whether or not to proceed and what he needs to do next. This advice should take account of the client's preferences with respect to his own responsibility for, and involvement in, the risks of his own project. He needs to understand that the more he transfers the financial and managerial responsibility and risks to others, the greater the burden upon himself to define his requirements fully and accurately at an early stage in the process. Failure to do so will expose him to the real possibility that the building will not meet his requirements, even if it is built on time and within the contracted sum.

c)   Promoting an environment which insists upon good communications....... and encourages an integrated approach to the whole procurement process

The client has the right to expect that his project is well managed. By the same token this places an obligation on his own organisation to be well administered, and that information and decisions are co-ordinated and communicated clearly and consistently. It also means that the organisational arrangements and roles of all those taking part must be clearly defined, that contracts and conditions of appointment are integrated; and that channels of communication are set up in a way that will encourages good working relationships and a spirit of team working.

It is in the client's own interest to support initiatives that stimulate a holistic approach to design and construction and encourage the integration of the constructors in the design process. Positive involvement by the client can cut across artificial lines of demarcation and promote co-operative working practices.

c) Setting criteria for selection and appointment of advisers and contractors

The client's objective in selecting his advisors and builders is to find people who have the experience and expertise to undertake his project and the ability to work together, and with him, as a team. The client may elect to stand back from the detailed workings of the design and construction stages, but he cannot afford not to be involved in the selection of those who are to manage it for him. Nor can he stand back from the process of establishing the selection criteria, which must adequately define his requirements for the project. He must accept that it is his role to lead this task and that it requires thoroughness, fairness and considerable care.

d) Resourcing, through the provision of funding, information and other essentials including enabling through supportive authority

The client's vision is only achievable if he is prepared to support it fully. Providing adequate funding is only one, albeit crucial, aspect of the client's backing for his own project. The task of providing briefing information is just one of many that distract and disrupt, unless dedicated resources are made available. Any operation requiring the movement or relocation of personnel invariably creates tensions and difficulties which can only be resolved by management who have the time and authority to do so. Other resources can be brought in from outside, but it is often best if these can be found from within his own organisation. The client who opts for a hands on approach to the running of his project will have to take on additional tasks of co-ordination and management, which also require his direct authority.

e) Entering into contracts and accepting the responsibilities imposed by them

Decisions made about the method of procurement have to be backed by relevant contractual arrangements, which, amongst other things, define the duties and describe the roles of the key participants including that of the client.

Where contracts have been entered into with the client's full involvement there should be no surprises. Difficulties arise where role definitions have not been made clear, they conflict with one another, or simply if the client is unaware that the contract invests some of his authority in his agents. For example, consider the predicament of a client who is unwilling to pay for a finished piece of work because he does not like it, even though the contract administrator certifies that it meets the requirements specified. In such a situation it is largely immaterial whether the fault lies with the client, who has either omitted to find out what he is getting, or has failed to communicate his intentions clearly enough; or whether it lies with his advisers who have failed to interpret his needs correctly and then inadequately explained what was specified. What matters is that such a situation could have been avoided if all parties had taken the trouble to be aware of what was specified in the work and the context in which it would be administered.

f) Reviewing performance against targets - and taking action if not met

Although the client agrees specific targets with those he employs, he would be foolish to assume that this was the end of the story. One of the less endearing, and for clients, frustrating features of the construction industry is its optimism about its own performance, despite repeated failure. Past experience fails to dim this confidence which often remains unaffected by criticism and poor performance.

The experienced client knows that regular pressure applied through the process of reviewing the project targets, particularly those relating to time and cost, pays off in a number of ways. First, it serves to emphasise the value the client attaches to ensuring that targets are taken seriously and adhered to. Secondly, it encourages a regular dialogue and understanding about the elements of the project subject to risk, and gives the client the opportunity to share in the making of contingency plans to overcome difficulties.

Although the client usually has a contractual basis for the goods and services he requires, including the standards and targets, there are few procurement arrangements where the client does not bear some measure of risk: even a Package Deal may involve risk. Promoting or resorting to an adversarial approach when things have already started to go wrong is unlikely to bring the client any real benefits, except in the most extreme circumstances where mitigation of loss is the best that can be hoped for.

The category B tasks

These tasks are principally concerned with supporting and administering the client's role as enabler and leader. They may be delegated in totality to the client's representative, or shared between the client's representative and the project manager/ lead consultant. The choice of procurement path is the primary factor that determines what needs to done and who should do it. A singlepoint arrangement, for example reduces the client's involvement in the building and design processes but may well increase the need for internal project management. In deciding how to set up his project the client has to be aware of the importance of the need for clarity of roles and independence of action by those who represent his interests. Whatever the degree of delegation, it is the client who must retain overall responsibility for ensuring that his project is properly managed and co-ordinated

The Client as a Force for Change

Influence - for better or worse

The greater the skill and experience, the greater potential the client has to influence building practices and procurement. However, the combination of skill and experience is not a guaranteed recipe for success. A brief examination of the role of some major clients of recent years provides a useful insight into the way in which their actions have influenced, reinforced and shaped the construction industry's procurement practices as well as contributing to the success or failure of their own projects.

The public sector experience

Until recently, local government, particularly housing authorities, commissioned a large slice of the industry's workload. Most of the bigger authorities directly employed their own expert workforce, both to assist in the management of their building programmes and carry out some or most of their work. The remainder was shared out and commissioned to at least an equal number of private consultants. The majority of the construction work was undertaken by private contractors. The whole process therefore included a fair cross section of the industry. The size of many of these authorities was such that they were able to establish their own design departments with their own standards, detailed briefing arrangements, and in some cases R & D. The LCC, as it then was, the GLC, PSA and others were regarded in their heyday to be centres of excellence for construction, management and design. So what went wrong? Why did so many contracts overspend and over run? Why did so many buildings fail in both planning and construction? Was it the designer, the builder or the client who was to blame? A look at what was happening at the time gives some clues.

The deficiencies of existing housing provision in the early 1960s is well documented. It nevertheless provides an important and powerful context for the actions of the clients of the time: the immensity of the political will to build in large numbers over short timescales. The realisation that the demands placed upon the building industry's resources would not be met by traditional building methods alone lead to the positive encouragement of factory production techniques, system building and rationalisation of existing ways of working. This already difficult if not impossible task was further complicated by the intransigent requirement to build at ever higher densities.

The political imperatives for speed and quantity outweighed any misgivings. It was however acknowledged by some at the time that the adopted solutions carried inherent risks. Those that recognised the shortcomings of some of the more 'popular solutions' of high rise, prefabrication, etc., were faced with the almost equally bleak prospect of continuing to stretch the industry's comparatively slender resources still further by resorting to more traditional forms of construction. Faced with a similar situation again, would the client's demands be tempered and be brought more in line with available resources; and would the industry again respond with inadequate and underdeveloped design and technology? Even if the technology had been available, what could or should have been the response to the demand that projects were underway in timescales that failed to allow sufficiently for full preparation of documentation? Despite advice and warnings to the contrary huge numbers of lump sum contracts were let on the basis of scant or incomplete information.

When the clients' demands stretch the industry's resources to the point where it becomes impossible to put together a satisfactory tender list of six suitable contractors, procedural safeguards such as 'approved lists of contractors' lose their significance and value, particularly as reliable instruments of accountability. The choice was real but stark: don't build, or get on with it and accept the consequences. Visible accountability is one of the major arguments put forward in favour of competitive tendering where price is the single, if not sole element of competition. Clients, particularly public ones are much more reluctant to engage in methods that require more complex, less visible and often more difficult forms of assessment. This attitude, together with an inadequate insistence upon readiness of information encouraged an increasingly claims orientated and adversarial approach to building procurement, where penalties not rewards underpin the arrangements.

Influence to stagnate?

This adversarial approach has been further reinforced, often with the best of motives, by the 'experts', often technical professionals, who, as well as representing their client's attitudes, have tended to add their own views to the working parties and committees responsible for drafting and updating the industry's most commonly used standard forms of contract. It is not surprising, therefore, that the traditional forms of contract can seem positively to discourage any integration of the design and construction process.

Clients in the driving seat

Dissatisfaction with the practices of traditional procurement has lead the more experienced and independent clients to adopt systems that cope more effectively with their increasingly large, complex and technically advanced projects. Mostly these have built upon experience gained from the United States and some large organisations who have exacting technical needs, such as those within the petrochemical industry. The need to manage and control the construction of highly sophisticated technical installations has compelled designers of all kinds to work more closely with the constructors. Frequently it has been the clients themselves who have been the driving force behind this innovation. By employing their own experts to help manage their projects, they have been able to concentrate first and foremost upon selecting and building teams with both the expertise and willingness to work closely together, each contributing within a single management framework. Common objectives are backed by an attitude that rewards success but deals firmly with obstruction or inadequacy. Whether the process is managed directly by the client's own organisation, or through a carefully selected agency, it is the client's insistence upon and willingness to pay for fully integrated working that can enable intricate construction to be achieved within tight targets and quality standards.

It is not just clients with complex building requirements who have pressed for change. The need to control less technically demanding projects more precisely, particularly with respect to time and cost, has lead to a number of initiatives over the last fifteen years for change and improvement. These have come from individual clients with the purchasing power to develop their own approach to the procurement process, as well as influential client bodies such as the BPF with their system aimed more specifically at the medium sized project. Many attempts have been made to set clearer targets and to cut through some of the more divisive working practices by eliminating waste in both time and effort. Such systems rely heavily upon the client's own ability to understand and manage the process; they also assume that the industry is able to provide a relatively high level of expertise across the whole spectrum of its operations and particularly in relation to providing a meaningful input into the design process. It is my experience that many medium sized contractors do not have the inclination, or have not yet acquired the skill, to work in these ways.

Design and build

The recent increase in the use of design and build, particularly by clients who have a regular development workload but who wish to stay at arms length from the process itself, is a further indication that the client is dissatisfied with the unreliability of delivery by other methods. From the client's point of view, however, design and build not only shifts much of the responsibility for risk onto the design and build organisation, but also encourages the industry to make buildability a more important priority than it does under the more traditional arrangements. The client, nevertheless, is perhaps not as free from responsibility, or from risk, as he may care to think. The tasks of establishing an accurate brief at a very early stage, and then making the right selection of the design and build organisation, become even more crucial if he is not to be disappointed by the end result: risk cuts both ways.

Some Key Issues

Although the client's position is central to the procurement process, his effectiveness is heavily dependent upon his ability to:

- understand the options, and processes in which he is required to take part
- make informed choices about advisers and contractors
- communicate his needs to the industry as a whole
- take some responsibility for the way in which the industry develops

Starting on the right lines

The first-time client accounts for a significant proportion of the construction industry's workload. How can the newcomer get started with the confidence of knowing that he has considered all the strategic options open to him? He can really only do this by taking informed, impartial and independent advice before committing himself to a particular procurement route. The danger is that the inexperienced are likely to turn first to friends and colleagues and may be encouraged to take a course of action unsuitable for the project that they have in mind. Even when the prospective client seeks professional help, there can be no guarantee that the advice received will be sufficiently impartial, unless it is given on a firm 'no commitment' basis. It is also unrealistic to assume that the professional institutions, who provide client advisory services, are any more likely to offer completely unbiased advice, especially if it runs contrary to the interests of its members.

A number of less partisan organisations, such as CIRIA, with its new series of Client Guides, have started to turn their attention to the client market. At present these client services are relatively limited and tend to be London-based. To be really effective such guides must be easily accessible across the country and sensitive to local as well as national issues. What seems to be needed is the construction industry's equivalent of the Citizens Advice Bureaux. A network of independent advice, provided by experienced construction experts, drawn from a range of different backgrounds, specially trained, accredited and registered by the construction industry. An arrangement on these lines could not only be of benefit to the industry's clients, but could also be a useful vehicle for feed-back and dissemination of good practice. After all, both sides of the industry would stand to gain from clients who are better informed.

Finding the best basis for selection

Historically, clients have approached the task of selecting their advisers and contractors in different ways. Until relatively recently, most of the professional bodies operated a system of mandatory fee scales. Whatever the merits of this system, it had the effect of relieving the client of the need to examine the price of the services he required, thus allowing him to focus instead on the quality of skills and expertise on offer. By contrast the selection of the building contractor most frequently used price as the major, if not exclusive, criterion for selection. Pressure to abandon fixed fee scales on the one hand, and a realisation on the other, that price is only one of a number of important components in the selection, both advisers and contractors has further complicated the client's task.

The conundrum for the client is that he has to decide before he receives the services that they will meet the standards he has in mind, and at a price that he is prepared to pay. If he has no direct experience of the people or organisations offering the service he is obliged to rely on information from others about past performance. This information is only likely to be of use if the tasks and context used for comparison are truly similar, that is, all the main ingredients, including the participants, are the same. Reliable information of this sort is often hard to come by.

The importance to the client, especially the newcomer, of making informed choices cannot be overstated. This is particularly true for the public client whose actions are not only open to public scrutiny, but subject to both external legislation and internal standing orders which prescribe the tendering process. The requirements that these impose can have the effect of limiting initiative and innovation and encouraging the use of existing systems for the sake of familiarity. Even the experienced client can, therefore, benefit from help and guidance on good practice in this difficult area. Approved lists and similar sieves cannot be expected to contain all the data: the most vital material is often the most sensitive. Such tools must be regarded only as a first step in a much more focused examination of performance in the selection process.

Therefore it is not surprising that clients with regular and repetitive building programmes tend to favour arrangements with limited number of teams who have demonstrated their ability to deliver. In such cases the client may confidently place a higher premium on his priorities for reliability and quality than he can when the service standards are unproven. The only measurable difference between one organisation and another is price. Recognition by Banwell[3] and now by *Constructing the Team* that partnering arrangements can be regarded as sound, provided that they have been properly arrived at, must be regarded as a sign of encouragement for those who genuinely support a move toward a team-orientated approach to building procurement

Meeting the clients' needs

Many clients see themselves as too small to have sufficient clout to change, and thus improve, the construction industry's services. They do not have the purchasing power to command the same authority as Marks and Spencers or BAA, nor do they have the expertise that goes with it. The clients' concerns for quality and value for money are clearly acknowledged in *Constructing the Team*, which is justifiably critical of the many obstacles that the construction industry has put in the way of improving it's own performance; it recognises the potential disadvantages as well as opportunities that could result from the fragmentation of government departments and other public agencies; recommends that 'Government should commit itself to being best practice client'; and proposes the creation of 'A Clients Construction Forum' to represent private sector clients. Whilst such proposals, if implemented, could indeed provide a useful client group to bring pressure for better information and standards, it must also be recognised that the construction industry itself has a

---

3 Banwell, H (1964) *The placing and management of contracts for building and civil engineering works.* HMSO; London

responsibility to ensure that the services it provides really do meet the clients needs.

It would be unfair to suggest that no attempts are being made by the industry to improve its standards and provide more information about its practices. The *Building Britain 2001*[4] report and others[5] have highlighted a number of initiatives intended both to improve the way the industry works and assist the client in his quest for feedback and information about the industry. *Constructing the Team* proposes that 'a register based upon ConReg should be compiled of consultant firms seeking public sector work in the United Kingdom.....'. Whilst initiatives of this sort must be welcomed, the practicalities of maintaining such a records as a useful tool are fraught with problems, as those who have tried to do this can testify. The difficulty the construction industry has in providing meaningful benchmarks about its performance is exacerbated by the proliferation of procurement options, and by the fact that such procurement options frequently involve the combination of work by several separate organisations. It is, therefore, often difficult to establish the worth of the individual contribution, particularly in the case where things have gone wrong. Despite this it has to be in the industry's long term interests to find out what the client needs to know, establish benchmarks and provide as much information about them as possible: these are essential if the client is to be able to fulfil his role effectively.

The responsibility of influence

The client as patron has a responsibility, as well as an interest, to ensure the continued existence and health of the construction

---

4   *Building Britain 2001*, Centre for Strategic Studies in Construction , Reading, 1988.

5   *Investing in Building 2001*, Centre for Strategic Studies in Construction , Reading, 1989.

industry. Greater complexity and increasing dependence on new building technologies brings corresponding increases in the necessity for sound management and good construction skills. New building techniques inevitably bring changes to the processes of designing and building. The client has to be aware and receptive to change and be prepared to adapt his role accordingly.

Above all, the bigger the client, as user and sponsor of the building industry, the greater his responsibility for the consequences of 'stop/ boom' policies. Neither stop nor boom allow the industry to train and retain the skills so desperately required at all its levels. The client should be the first to recognise that the effects of over-employment are almost as disastrous as under-employment, and that both almost inevitably lead to failure and dissatisfaction. It would be encouraging to think that if the government were to accept the challenge of committing itself to 'best practice client', as recommended by *Constructing the Team*, it would see the sensible regulation of its own spending on building brought into line with the industry's ability to deliver, as one of the cornerstones of its approach to improving its own role as a major client to the industry.

Postscript

The choice of procurement path affects the role and level of involvement that the client has in his own project, but it does not remove the basic obligation to undertake certain fundamental tasks:

- The client's single most important task is the selection of his advisers and agents.
- Enabling and encouraging a spirit of co-operation and teamwork is often more important than the choice of procurement path

Evidence suggests that where a client is prepared to get involved in a positive and informed way the procurement process benefits. The converse is also true. Failure to be sufficiently involved and understand the process can result in setting unachievable targets which almost inevitably lead to misunderstanding and confusion.

The client can, and does, achieve changes for the better by both individual and collective influence and pressure. The achievement of consistently good standards is unlikely to be an industry norm unless the client both demands and pays for them. The variable standards of the United Kingdom construction industry are in part a reflection of the client's knowledge and expectations of it. The industry's wholehearted willingness to listen to what the client wants and provide better information about services and standards can only benefit both the industry and its users.

9. Risk, construction contracts and construction disputes

Michael O'Reilly*

Synopsis

This paper considers research in the field of risk assessment and management as it applies to the construction sector and outline specific areas of application for risk assessment and management in the context of construction contracts and disputes.

Introduction

Risk management has in the last decade established itself as an important element in the construction manager's toolkit[1]. Several factors may be suggested to explain why it has been accepted so readily. First, the need to develop a 'risk-balanced' portfolio of work was highlighted by the failures of developers who overstretched themselves in the late 1980s. Secondly, the rising popularity of non-traditional contract and funding arrangements, for example; design-and-build, build-operate-transfer, privately-financed infrastructure project, has presented new types of risks and opportunities which need to be evaluated and assessed.

---

[*] The author wishes to thank Dr Michael Mawdesley of the University of Nottingham with whom he has had many conversations on the subject of risk management which have influenced the content of this paper.
[1] For example, see Flanagan R. and Norman G., *Risk management and construction*, Blackwell Science, Oxford, 1993; Hayes R.W. and Perry J.G. (1985), *Risk and its management in construction projects*, Proc. I.C.E., Part 1, Volume 77; Whyte I.L. and Peacock W.S. (1992), *Site investigation and risk analysis*, Proc. I.C.E., May 1992.

Thirdly, researchers in construction management have latched onto risk as providing a framework which is conceptually elegant and intellectually satisfying. Most researchers in construction management have focused on the use of risk management in the context of project evaluation and the management of construction works. But the utility of risk assessment and management in the field of construction law and dispute procedures has also been recognised.

The process of risk management

Risk management is not a technique. It is a framework within which potential courses of action may be judged in terms of the risks and opportunities which they present, hence enabling the most appropriate course of action to be taken. Risk management involves a series of stages. These are most easily presented in the form of a flowchart:

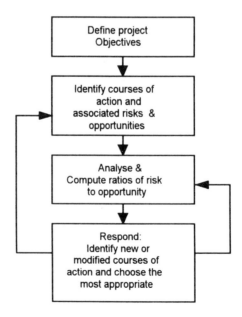

Figure 1 The process of risk management

In performing a risk management exercise in relation to any project, the first step is to define the objectives which the project is designed to achieve. Risk management iS not concerned with abstract risks, but only with risks which affect the attainment of defined objectives. Hence the proper and thoughtful identification of project objectives is a *sine qua non* of any meaningful risk management exercise. The second stage involves the identification of one or more courses of action which may achieve, in whole or in part, the defined objectives. Each course of action will contain risks and opportunities: risks are those factors which potentially threaten the attainment of our objectives, while opportunities are those factors which promote their attainment. A factor may of course be a risk as regards one objective and an opportunity in regard to another. In any venture the key function of a manager is to maximise opportunity while minimising risk, so a consideration of both is essential.

The third stage involves 'risk analysis'; we compute the levels of risk and opportunity associated with each of the potential courses of action which we have identified. Such an analysis may simply involve assessments of risk and opportunity based on experience. On the other hand we may carry out sophisticated computer simulations of projects which enable us to study combinations of circumstances. During this analysis, we compute 'ratios' of opportunity to risk. This requires the use of a balance (metaphorical or mathematical) in which the pros and cons of each identified course of action can be weighed. Clearly a course of action which is laden with opportunity but light on risks is attractive. For this purpose, the magnitude of a risk is derived by the multiplying the 'expected negative consequences in the event that the risk factor materialises' by 'the likelihood of its occurrence'. The magnitude of opportunity is derived in an identical way except that we concentrate on the positive consequences which may arise from uncertain events. Thus, an opportunity is no more or less than a negative risk. The concept of a 'ratio' must not be taken too strictly since many of the risks and opportunities cannot be directly expressed in the same currency. For example, in a construction project, the identified risks and opportunities may include 'money', 'time', 'personal safety', 'kudos' and others which cannot be directly measured using the same units of currency. In order to simplify our analyses we tend in practice to convert all risks and opportunities into 'money equivalent' terms, but there is no reason why we should not treat them separately in a 'multi-dimensional risk space'.

The final stage involves a response to the analyses which have been carried out. This may simply be the selection of the course of action which offers the most promising ratio of opportunity to risk. But frequently it involves a review of the lessons learnt during the previous stages; if we can devise a course of action which has hitherto eluded us we may repeat the analysis process in order to compare the balance of risks and opportunities which it provides with those expected from the courses of action which have already been identified. The process is, or should be, an iterative one. It involves a learning process through which we may begin to understand the way in which the factors associated with each candidate course of action threaten or promote our objectives. This may suggest to us new possible courses of action; these may be radically different from what has been considered already or merely modifications of them.

The benefits of risk management do not lie solely in the fact that we are able to select a course of action based on rational principles. Equally important, perhaps, is the fact that the process provides a structured and disciplined framework for thinking about the projects which we are considering undertaking. It does not in any way belittle the role of intuition and experience; it merely challenges us to justify our 'gut feelings' and emphasises the role of critical and rigorous thought. And through such a process it enables us to learn from our mistakes before we make them.

Risk objectives and perspectives

Risks, as we have already seen, do not exist in the abstract. They are risks because, and only because, they create a risk in respect of an objective which we have. A factor may thus be a risk for some and an opportunity for others. Take, for instance, poor ground conditions. Such conditions may substantially increase the cost of construction works. But where the contractor is being paid on a cost-plus-a-percentage basis, poor ground does not constitute a risk to him; rather it provides an excellent opportunity to increase his income. Even within the same organisation, objectives and perspective must be emphasised. Later in this paper we shall examine the use of risk analyses in the management of disputes. The perspective adopted in these simple examples is that of the lawyer/manager who is managing the dispute 'project'; it will be assumed in those analyses (for the sake of simplicity) that the lawyer/manager is concerned only to maximise the outcome of that dispute, without reference to external factors. In the case of the claimant this means to maximise the net recovery, and in the case of the respondent this means to minimise the net payout. But the organisation on whose behalf the lawyer/manager acts may not consider that such a perspective is appropriate.

For instance, imagine that the lawyer/manager acts on behalf of a contractor who undertakes a design and build project and who subcontracts out (a) the steel supply and erection and (b) the design services. Suppose that the steel frame is one of the earliest activities on site and that the design-build contractor is sued by the supplier of steelwork for failure to supply fabrication details on time. The supply of these details was the responsibility of the design subcontractor and any delay is likely to have been the designer's fault. Let us examine the problem from three different perspectives:

- that of the lawyer/manager who is managing the action commenced by the sub-contractor;

- that of the design-build contractor's project manager;
- that of the design-build contractor's managing director.

Upon receiving the steel subcontractor's writ, the lawyer's instinctive reaction may be to bring the design consultant in as a third party; that may be indeed be the 'right' response[2] when the success of the litigation, considered in isolation, is the chosen perspective. But when the overall objectives of the construction project are considered from the project manager's perspective, it may not be sensible to upset the design subcontractor at this stage by serving third party proceedings upon him; his continued goodwill and flexibility may be crucial to the timely completion of the rest of the project. Hence, the correct decision when viewed from the project manager's perspective may be to defend against the steelwork subcontractor and leave any action against the designer until after the works are completed. When viewed from the managing director's perspective, however, the most appropriate decision may be to let the subcontract designer off the hook altogether. He may, for example, be a 'friendly designer' who is engaged on a number of projects, who is extremely flexible and generally reliable despite his alleged shortcomings on this occasion. If the claim is small (say £50 000) and the designer carries out design works for the design-build contractor with a value of £5M a year, it may be detrimental to the overall long-term objectives of the contractor to involve him in the action at all. This analysis illustrates the point that we have to be extremely careful when performing risk analyses to ensure that we are approaching the problem from the proper perspective.

In terms of Figure 1, the question of perspective is dealt with in the first activity, 'define project objectives'. In order to do this, we must begin by defining the project. In the example discussed

---

[2]   I.e. the response which maximises the ratio of opportunity to risk.

above, this may be restated in the following terms: is the 'project' about which we should be concerned (a) the dispute, (b) the construction project, or (c) the continued overall success of the client? It would, of course, be unusual for (c) not to be the proper 'project'. Yet it is surprising how many managers fail to consider the question of perspective.

Risk and construction contracts

One of the principal functions of a contract is to distribute risks and opportunities between the parties and hence risk is an essential consideration in choosing contract strategies and in the drafting of contracts.

Contract strategy

The choice of contract or procurement strategy can be understood in terms of risk. On the face of it, different procurement systems are mainly concerned with matters such as design responsibilities and payment schemes. These, however, are all interrelated by their bearing on the risk allocation within a project. For instance, the acceptance of 'design responsibility' by a contractor affects not only the provision of design resources but presents risks and opportunities: there is a risk of design liability and an opportunity to design a structure which can be built more quickly and cheaply.

Some commentators have correlated the degree of risk allocation between the parties with the type of contract. For example, Treasury Guidance[3] suggests that under 'Design and Build' arrangements the majority of the risk rests with the contractor; under traditional pre-planned contracts there is a more balanced

---

[3] Central Unit on Procurement , *Managing risk and contingency for works projects*, Guidance 41, H.M. Treasury, August 1993.

allocation of risk; and when using 'Management Contract' arrangements, the employer carries the bulk of the risk. This analysis is sound, but incomplete because it fails to account for non-financial objectives. The Treasury analysis relates principally to financial risks, that is risks to financial objectives. If fast completion were the prime objective, for example, if we were involved with an urgent military project, the risk associated with each type of contract strategy may be quite different from that derived from a financial risk analysis. Hence it is crucial for those advising parties on appropriate contract and procurement strategies to do so with a clear understanding of the objectives of those parties.

Contract draftsmanship: general comment

It goes without saying that attention should be paid to the clear drafting of contracts. Uncertainty as to the meanings of contract terms reduces the effectiveness of project management as resources need to be channelled into discussions about the division of responsibility within the project. Ultimately, uncertainty may lead to conflict[4]. Conflict leads to expenditure on legal services. Imprecise contract provisions produce a risk that third parties (lawyers, expert witnesses, arbitrators) will become entitled to a slice of the cake which the parties thought that they were going to be able to divide up between themselves.

---

[4] This was amply demonstrated in the recent litigation in *Linden Gardens Trust Ltd. v Lenesta Sludge Disposals Ltd.* and *St Martin's Property Corporation Ltd. v Sir Robert McAlpine and Sons Ltd.* [1993] 3 All. E.R. 417. Ian Duncan Wallace QC cites the *Linden Gardens* litigation as an example of the consequences of the 'lamentable quality and lack of precision of typical standard form draftsmanship': (1993) L.Q.R. 82 at 91.

Drafting specific terms

The risk allocation function of a contract is accomplished by writing terms which control risk situations, such as design responsibility, potential poor ground conditions, inclement weather and price fluctuations. Where such matters are not dealt with expressly, the law takes a view on the allocation of risk. A well-known example is to be found in Bottoms v Mayor of York[5]. In this case the contract was silent as to whether the contractor was entitled to extra payment for difficulties due to unanticipated ground conditions; the court decided that the occurrence of such conditions was at the contractor's risk and there should be no extra payment. The general principle in contract law is that commercial parties may make, within very broad parameters, whatever agreement they wish. They may allocate risk as they like. However, it is widely recognised that projects run more smoothly and efficiently with a lower expected cost and time if risks are allocated appropriately. Hence, we may identify a 'code of good practice', which provides appropriate solutions to the question of how risks should be allocated if the aggregate commercial advantage of all those involved is to be maximised.

A number of principles have be suggested. These have been expressed in a variety of ways and different writers have emphasised different principles[6]. However, there is an

---

[5] (1892) Hudson's Building Contracts, 4th Edition Vol. II p.208

[6] See: Abrahamson, M.W., *Risk management*, [1984] ICLR 241, Barnes, M., *How to allocate risk in construction contracts*, International Journal of Project Management, Vol. 1 No. 1, February 1983; Chapman, C.B., Ward, S.C., and Curtis, B, *Risk theory for contracting*, Construction Contract Policy, Uff, J. and Capper, P, Editors, Centre of Construction Law and Management, King's College London, 1989; Dering C. *Risk sharing and contract forms: devising appropriate contract terms*, Future Directions in Construction Law, Centre of Construction Law and Management, King's College London, 1992; Perry J.G. and Hoare D.J. 1992, *Contracts of the future: risks and rewards*, Future Directions in

encouraging consensus, with which the present writer broadly agrees. The principles set out below attempt to summarise this consensus. In particular cases it may not be possible to satisfy all of them simultaneously, but a sensible, judicious balance may be struck.

- Risks should be identified and a conscious decision about managing each major risk should be taken. A variety of strategies are available for dealing with individual risks, including retention, transfer and sharing. When a strategy has been decided upon it should be expressed as a term in the contract.

- Allocation of risks should be clear, complete and unambiguous. The more significant the risk, the greater the need for clarity. Uncertainty about the meaning or ambit of a term can itself be a major source of risk.

- The allocation of risk should be 'motivational'. This means that its allocation to a particular party should have the effect of motivating that party to deal with it in the most effective and efficient way.

   This implies that the party who accepts the risk should be able to:
   - influence its magnitude;
   - control the effects of the risk once it has occurred;

Construction Law, Centre of Construction Law and Management, King's College London, 1992; Morris P.W.G., *Current Trends in the Organisation of Construction Projects*, Future Directions in Construction Law, Centre of Construction Law and Management, King's College London, 1992.

and should have an incentive for minimising and controlling the risk.

- If one party can shoulder the effects of the risk once it has occurred, but the other may not be able to do so, the more capable party should bear it. This is because in highly interactive contractual settings as occur for most construction contracts, significant risks being carried by one party also represent significant risks to the other. For instance, risk X may jeopardise the solvency of the contractor. But if the contractor fails, the employer will be left with an abandoned job. Hence, risk X is, by reflection, a risk to the employer also, and it is not in the employer's interest to place such a risk outside its own control.

These principles are designed to minimise the aggregate risk over the project. However, they are by no means sacrosanct and may be disregarded wholly or in part in order to tilt the balance of risk away from one party onto the other. Their status may be described as 'baseline principles', principles which represent a useful starting point for drafting contracts. As several authors have pointed out, it is often difficult to apply the principles in practice but their use reduces the likelihood of drafting based on the unwarranted prejudices of the draftsman. In practice, many employers may be reluctant to take on risks which can be passed to the contractor. However, sound risk assessment followed by an application of the above principles should produce the minimum aggregate risk and hence the most efficient and cheapest project, which is to the employer's benefit in the long run. And as the choice of contract basis is normally the employer's, the employer can use these principles to his advantage.

Risk allocation strategies

Once risk factors are identified, there are a range of ways of dealing with them including risk retention, transfer and sharing. These responses are briefly illustrated below.

a) risk retention

In cost-plus contracts the employer agrees to pay the contractor what the works have actually cost plus a fee. The employer deliberately retains the risk of increased construction costs[7]. This is often because he has important non-financial objectives; for example he may want early completion of the works. Since cost-plus arrangements permit overlapping design and construction and hence allow earlier completion dates, they may be attractive to the employer. In order to secure the opportunity of early completion, employers often deliberately retain the risk of cost overruns on the construction project itself.

b) risk transfer

Devices used for risk transfer include insurance, transferring design obligations (e.g. design and build), and subcontracting. Insurance is the classic device for transferring risk and the basics of risk management derive from the models of the insurance and reinsurance markets. The insured's risk is transferred to the insurer for a premium. The insurer takes on a large portfolio of risks and, based on the overall probabilities derived from historical data (or other methods of estimation), sets a premium which should, on balance, return a profit.

Another device for risk transfer is to subcontract work packages. For instance, a main contractor may sub-let the ground works

---

[7] Although in many cost-plus contracts, a target scheme is operated in order to minimise the effect of cost overruns. This produces a risk sharing arrangement.

package at a price adjustable if poor ground is encountered, but only payable if the main contractor's entitlement to that additional price is proved. In many supply contracts, there is also a marked element of risk transfer. For instance, many roadworks contractors require their bitumen/asphalt suppliers to set a price applicable for the duration of the work, irrespective of the fluctuations of the international oil markets. The suppliers thus buy the opportunity to supply for a major project but also take on the risk of a rise in the oil price. They charge a premium for the fixed price; from the contractor's perspective, the premium he has to pay is the cost of buying security.

c) risk sharing

A common risk sharing strategy involves the use of joint-ventures. An employer may stipulate the use of joint venture contractors to reduce the risk of contractor insolvency. From the joint venturers' perspective, the risks of making a loss are distributed between them, but the opportunities of making profits are diluted in the trade-off. Risk sharing arrangement are often found in the terms of construction contracts. The ICE 6th Edition, for instance, provides risk sharing arrangements for weather and ground conditions. In both cases, the contractor bears the full risk up until a certain point; if the conditions are worse than a specified limit, the employer bears the cost and time effects of the poor ground conditions and the time consequences of exceptionally adverse weather.

Risk considerations in dispute management and resolution

Whenever lawyers commence the process of litigation or arbitration, they see the preliminary evidence and can form a provisional view on the prospects[8] . Likewise they can form a sensible assessment of the likely costs. Such assessments are traditionally communicated to the client in rather vague terms such as 'good prospects' or 'a fairly expensive action, perhaps running to six-figures'. Proper quantified assessments of likely costs and recoveries are, however, key ingredients in risk assessment. Construction lawyers must therefore develop the skills to provide quantified assessments.

Simple calculations of risk

Consider two examples where risk principles can assist in making decisions about whether or not to pursue a dispute. These examples are rather contrived, but they illustrate the technique. More sophisticated and practical analyses can be performed using computer simulations[9].

a) Example 1

A contractor is in dispute with his employer on a 'ICE 6th Clause 12' matter. It is a case which will be decided one way or the other depending on the arbitrator's view as to whether or not 'conditions could reasonably have been foreseen by an experienced contractor.' The contractor is advised by his experts that the prospects are 50-50, that the amount in dispute is realistically valued at £50,000 and that if he is successful in his claim his costs will be paid in full. His costs of running the case will be £35,000, the respondent's bill will come to a similar

---

8   Though the Heilbron Report (*Civil justice on trial: the case for change*, Report by the Independent Working Party set up jointly by the General Council of the Bar and the Law Society, June 1993.), Paragraph 4.1.(ii) suggests that in many instances, the clients' advisers pay insufficient attention to the evaluation of the dispute in the early stages, which leads to an increase in the overall cost.

9   See below.

amount and the arbitrator's costs will be £10,000. So if he is unsuccessful, he will be liable for a total costs bill of £80,000.

From this information we can compute that the outcome in 50% of cases will be +£50,000 and in the other 50% will be -£80,000. A simple calculation (0.5 x £50k + 0.5 x -£80k = -£15k[10]) indicates that the expected overall outcome is a loss of £15,000. Hence it is poor risk management to continue all the way to arbitration. This does not, of course, necessarily mean that the contractor should immediately abandon the claim. He should, however, make strenuous efforts to have it settled. The employer will if he agrees with the estimates, be exposed to even greater risk and hence will have an even greater incentive to settle. The expected overall outcome from the employer's perspective is a loss of £65,000 (0.5 x £0 + 0.5 x (-£80k - £50k) = -£65k). This information should prompt the employer to make a substantial offer. So, if the parties are between them destined to lose £80,000 on average, where does that money go? Answer: to the lawyers, expert witnesses and others who advise the parties; they are the only risk-free players in the game.

b) Example 2

Contractor C is claiming £50,000 for additional works associated with a building project. The employer argues that the works for which the claim is made formed part of the original scope of works. C's advisers tell him that his case is good and that he may recover up to £45,000 with a more realistic estimate being £25,000. The employer has not made an offer to affect costs and C's advisers tell C that he may proceed without concern. However, C remains sceptical. He consults with his lawyers and

---

[10] This value is termed by managers and analysts the 'expected monetary value' (e.g. Moore, P.G. and Thomas, H., *The anatomy of decisions*, Penguin Business Books, 1988), or EMV. The EMV is the sum of all possible incoming and outgoing packets of money, each multiplied by the likelihood of its occurrence.

quantity surveyors and asks for a breakdown of the likelihood of recovering various amounts. He also calls for estimates of likely costs. The lawyers and quantity surveyors provide the following assessment of outcomes:

Recovery (£'000s)	Prospects
£ 0 - 10	10%
£ 10 - 20	25%
£ 20 - 30	40%
£ 30 - 40	20%
£ 40 - 50	5%

C's advisers further estimate that their costs of running the dispute all the way to a hearing will be £60,000 and that the employer's costs will be of similar magnitude. They reckon that the court will allow the successful party to recover about two-thirds of the costs it has actually incurred. A simple calculation[11] reveals that the expected outcome on the award is:

(£5k x 0.1) + (£15k x 0.25) + (£25k x 0.4) + (£35k x 0.2) + (£45 x 0.05) = £23,500.

Since the irrecoverable element of costs will be £20,000, the average net outcome will be just £3,500. Alongside this rather paltry expected recovery are the opportunities both of a greater recovery (up to £25,000) and, of course, a risk of making a substantial loss of up to about £15,000. So the original advice that there is no risk is clearly incorrect. Overall, on a proper balancing of the costs, potential recoveries and likelihoods of success - not to mention the disruption caused by the dispute - it

---

[11] In this calculation, the following simplification is used. For each band of recovery (e.g. £10,000 to £20,000), it is assumed that all recoveries are concentrated at the mid-point in the range (i.e. £15,000 in this case).

would seem inappropriate to take this case to a full hearing. An immediate capitulation may not be the best approach either because the employer will be subjecting himself to even greater risks than the contractor by continuing and hence will have an incentive to make an offer.

Suppose now that the employer puts in an offer of £10,000, what is C's overall expected recovery? C has 10% chance of recovering less than the £10,000 offer. If he fails to beat the offer, his total costs bill will be £100,000; i.e. £60,000 for his own costs and £40,000 for the employer's recoverable costs. In the other 90% of cases, he will beat the offer and his costs bill will be £20,000, that being the irrecoverable element of his own costs. The average costs bill for C will thus be:

$$(\text{£20k} \times 0.9) + (\text{£100k} \times 0.1) = \text{£28,000}$$

Since his expected outcome on the award is only £23,500 and his expected costs bill is £28,000, the overall expected net outcome is a loss of £4,500. It should be emphasised that this overall loss is not just a possibility - it is the expected outcome. In short, C is more likely to make a loss on the dispute than a profit and the offer begins to look very attractive.

Litigants have always viewed potential action in terms of risk; but their advisers have traditionally not been able to provide them with quantified risk assessments. These are by no means easy to produce given the uncertainties such as whether or not witnesses will come up to proof, whether or not the court/arbitrator will concur with a line of reasoning and so on. But, it is the responsibility of advisers to do their best to provide such estimates and to perform risk assessments. The above analyses show that simple calculations can provide effective guides to action.

Computer simulations of risk

a) Activity projections

In many projects, such as many straightforward construction projects, the activities required to achieve the objectives are fairly self-evident, or are known with a fair degree of certainty. However, dispute resolution 'projects' tend to be highly uncertain as to their constituent activities. Figure 2 shows a coarse probabilistic network or flow diagram, illustrating the views of an adviser on the potential stages in a series of dispute processes under a contract. It is assumed that negotiation will be the first stage, with a possibility of conciliation and a further possibility of arbitration if the conciliation fails, reflecting the dispute resolution process provided for in, for example, the Institution of Civil Engineers 6th Edition Contract 1991. A working version of such a network will no doubt contain several activities in the stead of 'arbitration' to reflect the fact that there are a number of stages during the protracted arbitration process at which a settlement is likely. Furthermore, activities may in practice proceed in parallel; for instance, negotiation 'behind the scenes' is a common feature of many formal arbitrations.

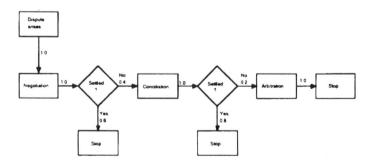

Figure 2 Simple probabilistic network for dispute resolution

In this paper we shall not explore the use of probabilistic networks, though a full analysis will require the issue of uncertain activities to be addressed. Details of techniques for the solution of probabilistic networks have been published elsewhere[12].

b) Time and cost information

Figure 3 shows projected activity times and costs for a typical hypothetical construction arbitration[13]. The times and costs are similar to what may be expected for a fairly typical small engineering or building project and this underscores the need to manage not only the underlying works projects but any dispute which arises from it.

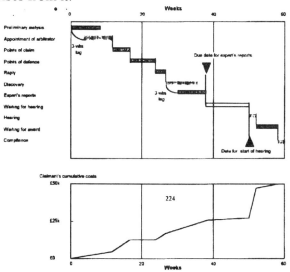

Figure 3 Cost and time information for an arbitration

[12] Elmaghraby S. E., *An algebra for the analysis of generalised networks*, Management Science, Vol. 10 1964, p.494
[13] This example was first published by O'Reilly and Mawdesley, *The evaluation of construction disputes: a risk approach*, Engineering, Construction and Architectural Management, Volume 1, No. 2, 1994. The author is grateful to Dr Mawdesley, Professor McCaffer (the Editor of ECAM) and Blackwell Science Press.

Although the links on the bar chart model are predominantly 'finish to start' connections[14], there is no difficulty in using precedence connections to model simultaneous progress on connected activities. Precedence start-to-start connections are used to link two pairs of adjacent activities. Two fixed dates are shown, indicated by black triangles; in UK arbitrations it is usual practice for important dates such as the exchange of expert witness reports and the hearing dates to be fixed at a meeting held shortly after the arbitrator's appointment. These fixed dates are occasionally re-set, but generally they provide a good control mechanism for the arbitration overall time. However, it is important to realise that they are set by the arbitrator in consultation with the parties and so at the early evaluation stage they will not be known.

Only one series of figures is given, perhaps representing the claimant's adviser's opinion of the most likely progress in terms of time and personal cost exposure for the claimant. But we can equally draw 'optimistic' or 'pessimistic' versions of Figure 3 in order to show the likely boundaries of outcome. This is useful for evaluation; the optimistic, pessimistic and most likely values of time and cost can be used to draw simple probability density diagrams for use in a Monte Carlo simulation[15]. For instance we may estimate that the quickest and slowest credible overall times for the arbitration are 40 and 100 weeks; and that the cheapest and most costly credible costs expenditures for the claimant are £38,000 and £65,000 respectively. We can represent this information using the 'claimant's cumulative costs' diagrams in Figure 4. In this diagram, the shortest credible time has been linked explicitly to the cheapest overall costs bill; and the longest time is linked to the most expensive costs bill. Such direct linkage is unlikely accurately to reflect the reality of the situation

---

[14] See any standard construction management textbook for the meaning of these terms and the theory and practice of network techniques.

[15] This is a technique for performing analyses on probabilistic data and is briefly explained further on in this section.

and it may well be that costs and time are relatively independent of one another. However, the format of Figure 4 does enable two important cases to be displayed in diagrammatic form which can be understood by clients and other advisers such as lawyers. Note that the extreme case in terms of rate of expenditure is that of maximum cost in the minimum time, which is not shown.

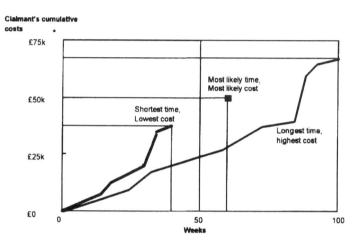

Figure 4 Cost information for the arbitration described in Figure 3 showing two combinations of upper and lower credible values of time and cost

c) Evaluation using probabilistic techniques

The cost and time information elicited can be use as the basis of a probabilistic evaluation of the risks and opportunities in the project. However, further information is required:

(i) an assessment of the likely principal sum recoverable if the case should proceed all the way to a hearing and award. This will probably be best described in terms of likelihoods of various ranges of recovery.

(ii) an estimate of the percentage of one's costs which will be recoverable in the event that one is 'successful' in the arbitration. Unless the successful party has acted unreasonably in the reference, the arbitrator will normally award him his reasonable costs, to be paid by the unsuccessful party. But rarely will the arbitrator agree that 'reasonable costs' are equivalent to the total costs bill which the successful client is obliged to pay his lawyers and advisers. It has been estimated that 65% of solicitor's costs tend to be recovered by the successful party, with a higher proportion of expert witness and counsel's fees[16]. The definition of 'successful' for this purpose is dealt with in (c) below.

(iii) whether or not an offer of settlement affecting costs has been made. This may radically affect the costs situation. A successful claimant is one who recovers more in the arbitration than the offer of settlement made by the respondent. A successful respondent is one who puts in an offer which exceeds the eventual recovery. The general principle is that a claimant who fails to beat an offer has to pay the respondent's costs from the time of the offer[17].

The estimates from (i), (ii) and (ii) above can be combined to form compound figures for decision-making. This may be achieved using probabilistic analysis, such as Programme Evaluation and Review Techniques (PERT) or Monte Carlo[18]

---

[16] Kendall, J., *Management of the dispute resolution process*, Future Directions in Construction Law, Centre of Construction Law and Management, King's College London, 1992.

[17] See generally O'Reilly, *Costs in Arbitration Proceedings*, Lloyds of London Press, Due April 1995.

[18] Monte Carlo techniques involve the simulation of the 'project' by selecting outcomes from the various sets of probabilistic data provided; they do this using random numbers (as suggested by a roulette wheel, hence the name of the technique) which operate on the data to provide a weighting towards the most likely outcomes in proportion to their

analysis. A program has been developed - CLARAT[19] - which is used to perform a wide range of probabilistic analyses. It prompts the user to input activities and estimates of costs and times using standard 'projects' which the user is invited to modify. It operates using Monte Carlo simulation and may present a range of graphical outputs.[20]

In the example under examination let us estimate that the principal sum to be recovered is estimated to lie in the range £50,000 to £200,000, with a most likely recovery of £125,000. This information may be crudely represented by the simple triangular distribution in Figure 5, where the recovery is expressed as a variable, x. This also shows an offer with a value of x_{offer}. In this case the offer is pitched at £80,000, giving a likelihood of failing to beat the offer of approximately 15% based on the assumed shape of the distribution[21].

---

likelihood. By performing a large number of simulations a picture of the likely distribution of outcomes can be obtained taking all the identified risks and opportunities into account.

[19] Claims, Litigation and Arbitration Risk Assessment Technique: See O'Reilly and Mawdesley, *Evaluation of construction disputes: a risk approach*, Engineering, Construction and Architectural Management Journal, Volume 1, No.2.

[20] Typically, CLARAT! uses 20 000 simulations, which a standard desktop computer can calculate and graph in 100 seconds

[21] The height of the distribution graph at any value of x represents the relative likelihood of the value x being the outcome. Thus the probability of the outcome lying in the range x1 to x2 is given by the area under the distribution which lies between x1 and x2 expressed as a proportion of the total area under the graph. In the present case, the shaded are on Figure 5 (which represents all outcomes less than the offer) is approximately 15% of the total area.

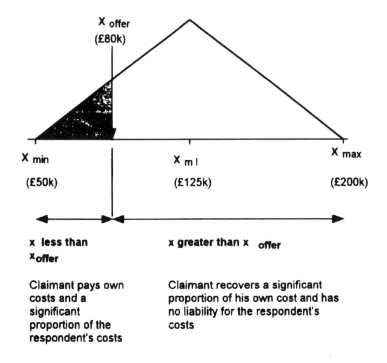

Figure 5 The assumed distribution of principal sum recoverable in the arbitration (x), shown with an offer of level x offer.

The proportion of costs recoverable upon success is estimated to lie in the range 50% to 100% with a most likely recovery of 75% (again a simple triangular distribution can be drawn). And for the sake of simplicity in this example it is assumed that the respondent's total cost bill is equal to that of the claimant. The boundaries of the feasible solutions are shown in Figure 6. This shows the 'binary' switch from 'being successful' (i.e. beating the offer) and being 'unsuccessful' (i.e. failing to beat the offer). If the claimant recovers £80,001, he is 'successful' according to this criterion. But he may still fail to recover up to 50% of his costs; if his costs bill is £65,000, he may thus need to apply up to £32,500 in paying his own lawyers and advisers, thereby reducing his net overall winnings to £47,501. This is the claimant's worst outcome for a 'successful' result. On the other hand, the claimant's best outcome assuming that it fails to beat the offer is a recovery of £79,999, with a personal costs bill of £38,000 and a liability to pay just 50% of the respondent's £38,000 costs bill. This leaves a net outcome of £22,999 [i.e. £79,999 - £38,000 - £19,000]. Corresponding figures can be prepared from the respondent's perspective.

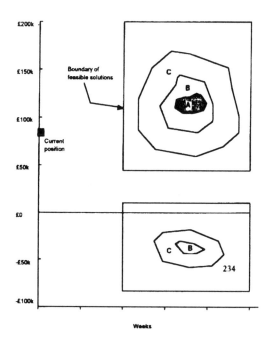

Figure 6 Contour diagram showing the typical anticipated spread of outcomes from the claimant's perspective

Using Monte Carlo simulation we can estimate the likelihood of particular outcomes within these feasible boundaries. We can plot the results of the Monte Carlo simulations using contours such as those illustrated in Figure 6. In this figure, Area A represents a high density outcome; B and C represent areas of increasingly unlikely outcome. This gives a more useful impression of the spread of outcomes expected than the boundaries of feasible solutions.

A useful direct way of comparing the claimant's and respondent's views is to draw probability density diagrams of the likely net financial outcomes relative to the current position if the offer were to be accepted, as shown in Figure 7. The position from the claimant's perspective is shown above the horizontal axis, that of the respondent being shown below. A study of the projected distribution of outcomes relative to the current position shows the risk experienced from the perspective of both parties, assuming that both agree on the estimates. The relative position of the parties can be gauged from the moment of the diagram around the current position (designated by a black circle on Figure 7): if the distributions represent intensity of load, a propensity to rotate clockwise indicates that the claimant is in an advantageous position relative to the respondent. If the moment indicates a propensity to rotate anti-clockwise, the position of the respondent is more favourable than that of the claimant.

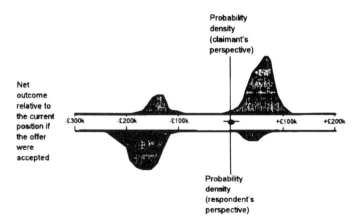

Figure 7 Probability density diagram showing the likely outcomes relative to the position of the current offer accepted - from the perspective both of the claimant and respondent

In this case, the respondent's position seems rather disconcerting and there seems to be good grounds for making an improved offer to induce an early settlement. The position seems more encouraging for the claimant, who will most likely improve his position. However, there are risks of a significant loss and, in any event, significant resources, funds and time will need to be expended to recover this improvement. Hence, there seem to be justifiable reasons for accepting the £80,000 offer, especially if the claimant is risk-averse. But many claimants in this position will not only look at their own situation but at the very great risk exposure of the respondent and will conclude that the pressure on the respondent will eventually be too great and that they will make an improved offer.

Put in general terms, what Figure 7 tells us is that the balance of risks weighs far heavier on the respondent than on the claimant and all other things being equal (e.g. both parties are financially sound, both employ sensible risk assessment advisers etc.) the respondent should increase his offer. A practical analysis will also need to recognise and account for the fact that both parties will form their own views of the likely costs and outcomes. Generally, the claimant will expect a higher recovery than anticipated by the respondent because parties tend to believe in their own case and, furthermore, both parties will not have access to the same information and evidence at the outset. Such a difference in view must be accounted for when attempting to predict how the other party will view its own risk exposure and hence how it will react. Differences in perception can be modelled by making estimates of the likely divergences in estimates between both parties and comparing the simulation results from computations with a range of divergences (a Monte Carlo simulation of Monte Carlo simulations).

Games theory can be of assistance in developing a risk strategy in such situations. Since arbitration, litigation etc. are games for two, or sometimes more, players, the objective risks which exist are insufficient to dictate appropriate strategy. There are many examples in this field of study, for example the prisoner's dilemma, which illustrate the need to take account not only of the perceptions of both players but also the perceptions of both players about the perceptions of the other and so on. In games theory, very simple examples create very complex solutions.

What if? analyses

One of the most useful features of simulation, whether performed by simple hand calculations or by computer, is the ability to carry out 'what if?' analyses. These involve running the simulation, then changing the assumptions to reflect possible changes in circumstances. For instance, we may vary the level of offer on the table. The distributions shown in Figures 6 and 7 can be compared for different levels of offer. As the offer increases, the risk shifts away from the respondent onto the claimant. Hopefully, by both parties understanding the risks, the offer will be pitched at a level which is 'risk - acceptable' to both parties.

Stage-by-stage analysis

So far, the discussion has proceeded on the basis that the relevant time at which the risk position is required is at the end of the project. However, since offers may be made at any stage, we need to understand how the attractiveness of any particular offer is affected by progress in the adventure. For instance, a low initial offer may be more attractive than a larger later offer because as time progresses more and more irrecoverable costs are expended. Using the 'What if?' process described, the attractiveness of different levels of offer at different stages can be investigated. This will enable a fuller risk strategy to be established in which acceptable settlement figures are determined for different times during the project. Such analyses may produce probability distributions of expended money and time at specified stages in the process.

Revised risk assessment during the proceedings

Risk analysis should not only be undertaken at the outset of a project, but at several stages throughout it. It is most usefully performed whenever there is a need to take a major decision. If the proceedings continue, it is important to review the situation from time to time during the process to understand the risks involved in continuing. One matter which has been touched upon already is that while the overall duration is very uncertain at the initial stages of a dispute, certain fixed dates tend to be established shortly after the arbitrator is appointed; these significantly reduce the uncertainty about the overall time and hence a revised analysis is useful at this stage. The analysis outlined above is equally applicable to any stage of the proceedings. As time progresses, of course, views on the level of costs, the likely recovery and so on will change and these changes should be incorporated into the revised model.

Design of dispute resolution systems

At the present time, there is much debate about appropriate systems for dispute resolution in the construction industry[22]. This paper will not to add to the debate on the relative advantages of the various proposals for reform. Instead, it seeks to highlight the utility of risk principles in the design of appropriate dispute resolution procedures and schemes. Any system, including a dispute resolution system, should be judged according to its success in promoting its objectives, enhancing opportunities and minimising risks to those objectives.

Consider construction arbitration as a case in point. Arbitration has attracted a good deal of criticism in recent years. It is said to

---

[22] For a recent survey, see *Dispute Resolution*, Construction Industry Council, January 1994.

be 'expensive, slow and nearly always left until the contract has been completed'[23]. However, much of this criticism is misdirected, as arbitration simply provides a framework for dispute resolution. If it is inefficient that is either because those who frame specific arbitration rules have provided opportunities for delays and excessive costs to build up or the arbitrators fail to control the proceedings, or both. Risk principles can be used to design an improved set of arbitration rules. First, the objectives need to identified. The following may be suggested:

- to produce 'the right answer';
- to produce a quick and timely answer ;
- to produce an inexpensive answer;
- to maintain good business relation between the parties; and
- to maintain confidentiality.

Clearly all of these cannot be guaranteed simultaneously and some compromise is required. For example, speeding up the process may create a risk for the objective of producing the right answer; giving time for full investigation of all matters will create a risk for the objective of producing a quick and timely answer. The overall design of the arbitration process can be approached by balancing all the objectives and the attendant risks and opportunities. The relative store we wish to place in each objective must be made explicit. For instance we may consider the right answer to be more important than the maintenance of confidentiality, and design the rules accordingly.

In the past, one objective - namely 'to produce the right answer' - has been emphasised, almost to the exclusion of the others. For example, arbitrations are held after the works are completed so that all the evidence is available, hence minimising the risk of

---

[23]   Sir Michael Latham, *Interim Report on Procurement in the Construction Industry*, December 1993

reaching the wrong answer; discovery by list has been traditional for the same reason; representation by counsel at hearings conducted along High Court lines assured that all the best arguments were deployed and tested, hence maximising the likelihood of reaching the correct answer. These procedures may have maximised the likelihood of getting to 'the right answer', but they have failed to give due and proper weight to the other objectives relating to time and cost. What is required is a more balanced approach.

Providing the rules of natural justice are adhered to,[24] the reported cases show that the law of arbitration is very robust. Parties may agree to have the arbitration conducted in a much shortened fashion, providing each party has a fair opportunity to present his case. This may be achieved using documents-only arbitrations or a 'chess clock' approach with each party being allotted a specified time for speeches, submissions and witnesses. In fact, there is little, except tradition and inertia, to prevent arbitrations being conducted quickly and cheaply, without creating undue risk to the objective of reaching the right answer. Risk principles show how a set of arbitration rules might be designed and provide a rational and intellectually-satisfying framework for the re-design. This article is not the place to embark upon a detailed analysis of the advantages and disadvantages of the solutions available; such an analysis requires that the 'objectives' be clearly defined and one of the problems - indeed, perhaps, the main problem - which currently exists is that there is no clear consensus as to the relative weight which the users of arbitration give to the objectives listed above.

---

[24]   It is submitted that Clause 10 of the draft Arbitration Bill (D.T.I., *A consultation paper on draft clauses and schedules of an Arbitration Bill*, February 1994), which is based on Article 18 of the UNCITRAL Model Law, expresses the principal, if not the only, requirement of an agreed arbitration scheme: 'The parties to a reference shall be treated with equality by the tribunal and each party shall be given a fair opportunity to present his case'.

Concluding remarks

This paper has focused on the use of risk management principles for the effective design both of construction contracts and dispute 'projects' and schemes. Risk principles are seen to provide a useful and flexible framework for thinking about such matters. They establish an intellectually consistent basis for balancing competing interests in the context of construction contracts and dispute resolution. By emphasising the objectives which are sought to be achieved and by defining risks and opportunities in relation to these objectives, more effective contract and dispute resolution strategies can be devised.

Part III

Particular Issues

10. Risk in Ground Engineering: A Framework for Assessment

Fin Jardine & Simon Johnson

Synopsis

This paper examines the various risks inherent in the ground and in the engineering of the ground and considers the implications for the management and procurement of construction works and services. The CIRIA programme of ground and contaminated land engineering research projects is described, giving guidance for reduction and management of risks in the ground.

Introduction

Lawyers can claim that their work provides the framework of society. All human actions, legal and illegal, in a civilised society have the context of law. Geotechnical engineers claim that their work is the foundation of society, everything being built in or on the ground. If lawyers are pillars of society, geotechnical engineering supports them. Unfortunately this is all too true. Only a tiny proportion of the activities of people lead to their going to law, though it seems an increasing trend, but quite a significant percentage, and again seemingly on the increase, of ground engineering activities lead to disputes, arbitration or litigation. Furthermore, very many of the claims and disputes in construction contracts have their grounds in a misunderstanding of the nature of the ground.

The commonest source of risk to construction projects is the ground; often it poses the greatest risk.[1] The ground (geology) is the prime source of uncertainty in geotechnical engineering.[2] The ground itself is usually innocent, but found to be guilty as accused, for only occasionally do its faults on their own cause failure. Rather it is the engineering that is at fault or inadequate, and that means that except for natural disasters such as flood and earthquake the blame, or, more fairly, the cause can be ascribed to human choices.

Origins of ground risks

The risks from the ground are the result of interactions between the ground and some agency in the context of time past, present and to come.

Ground	Nature and natural forces	Time
	People and human activities	
	The construction works and their engineering	

Thus, the voids of karstic limestone into which the ground above can collapse are the result of long, slow natural dissolution and erosion in the past; and land subsidence caused

---

[1] *Inadequate Site Investigation*, Institution of Civil Engineers, Thomas Telford, London, 1991.

[2] Muir Wood, A.M., *Control of uncertainty in geotechnical engineering*, Special Publication on Geotechnical Engineering, *International Conference*, International Society of Soil Mechanics and Foundation Engineering, New Delhi, 1994.

by brine extraction from salt deposits is the consequence of human activity; and the regeneration of a landslide on an ancient colluvial slope by an excavation for a new road is an interaction between new construction and the pre-existing environment.

Just as management and procurement are two-way processes, for example those between the manager and what is managed, so is risk. Thus, in the example of the new road, the works could regenerate or create a new landslide or, with prior recognition of the potential for instability, the design and layout of the road would include preventive works. Recognising the potential for risk is the critical step. But that is not enough: the wrong choice of preventive works could lead to a landslip. Stabilising the slope with a retaining wall might be a suitable option, but not if it dams the groundwater's natural flow and causes pore water pressures to rise, or if the excavation needed to build the wall is open for too long a time or over too great a length. While these are examples of poor decisions, it is the engineering works that create the conditions for failure. There are also many examples of work carried out to overcome one risk having a method or result that creates a different risk. A continuous flight auger can be used for forming a bearing pile through loose sand, but its action can both densify the adjacent sand or, like an Archimedean screw, pump sand from depth to the surface: in either case adjacent footings will settle or even be undermined.

In this country, we are fortunate in its temperate weather and its freedom from severe tectonic activity. We only consider the risks of the 'great forces of nature' for critical structures, when a rational and quantitative assessment of risk is made. In most cases the assessment, particularly of private clients, is qualitative, even subjective. People seem to accept the high risks of their garden slipping in to the sea, or of the tranquil river by which they live flooding their front room. Even when householders in Cornwall and Somerset were offered free testing for radon, only some 12% took advantage of this service.[3] For most the radon was an imposed risk not a matter of choice, because they did not know about its presence when they chose where to live. 'As safe as houses' is what we all expect, but the building must not move nor its plaster crack. But as nothing, other than being taxed until death, is 100% sure, the choices that people make always result in some 'small' degree of risk.

Definitions and preconditions

Blockley[4] has pointed out 'the need for settled definitions of basic terms in safety, risk and reliability theory'. He distinguishes between risk and hazard as follows:

> *'Risk is the combination of the chances of occurrence of an accident and hazard is a set of preconditions to the initiation of an accident sequence.'*

The latter definition (of hazard) is not always used. Sometimes people mean just one of the preconditions, methane for example, or they mean the hazard event, a methane explosion. In the definition given by Blockley, the hazard includes the methane, its migration route, its entry to a confined space, the ignition capability, and the persons or property endangered. Such a definition encompasses the contamination paradigm of: source – pathway – target, all having to be present to induce the hazard event.

---

3   *Householder's response to the radon risk: summary report*, Department of the Environment, HMSO, London, 1994.

4   Blockley, D.I., *Uncertain ground: on risk and reliability in geotechnical engineering*, Keynote Address to Conference on *Risk and reliability in ground engineering*, Institution of Civil Engineers, 11/12 November 1993, London.

This model applies generally to ground hazards, although there is interchangeability between the preconditions of the nature of the ground and its content, the ground engineering, the construction works and the people involved. In some cases the ground could provide the source and the pathway, or the people could make decisions which link the source to the pathway to the target, for example if stone columns in fill should channel surface water to depth and cause collapse settlement or if they act as vents to bring methane to below a building. The preconditions that give rise to a ground hazard can therefore be summarised as:

- the ground
- the content of the ground
- the actions on the ground
- the property on the site
- the people either present on the site or making decisions that affect the site.

Types of Ground Hazard

In Table 1, there is an arbitrary list of some potentially hazardous ground conditions. Note the innocence of features such as 'archaeology', 'time', 'animals', etc. Yet coming across a temple of Mithras in a basement excavation is likely to cause a long delay to the works and to cost the contractor, the consultant and the client dear.

Archaeology, artefacts, aquiclude, artesian conditions, active earth pressures, asphyxiation, animals

Bearing failure, boulder falls, blow, boils

Collapse settlement, corrosion, creep, compressibility, cracks, contamination

Differential settlement, dewatering, dissolution, driveability of piles

Erosion, earthquake, earth pressures, excavatability

Fire, foundations, fissuring, fill, flowslide, frost heave

Groundwater level change (rise and fall), gases

Heave, horizontal stresses

Instability. inflows, inrushes, ice lenses

Joints

Karst (solution features)

Loss of ground, landslips, leaching, liquefaction, landfill gas, landslides, lawsuits

Marsh gas, movement, moisture change, mudflows, methane

Negative skin friction

Obstructions to groundworks, organics

Pollution of groundwater, progressive failure, peat, piping, pore pressure

Quicksand, quagmires, quick clays

Rockfalls, residual soil, rupture

Subsidence, sinkholes, settlement, seepage, seismicity, swelling, shearing, shrinkage, solution

features, slides, slickensides, scour

Tension cracks, time, talus, tremors

Unsuitability for use (e.g. as fill), uplift, ultimate bearing capacity, underpinning

Vulcanicity (volcanoes), voids, vibration, volume changes

Weakening of support, water pressure, weathering

X - xenoliths (in other words, foreign bodies)

Yielding

Zone of shearing or yield

Table 1 Some potentially hazardous conditions of the ground

Environmental Risk

Types of risk and consequence

The main type of risk considered in this section is that of chemical contamination of the ground and groundwater. Contamination can occur naturally and arise from man's activities, both past and present, and may be a direct result of deliberate use or of accident. The contamination can be obvious: stagnant ponds, discoloration of the soil, noisome odours, drums leaking noxious liquid, etc. On the other hand, the contamination can be difficult to identify, and the source of the contamination or the contaminants themselves might be spread over a wide area. It could be in 'hot spots' or dispersed, or it might be dissolved in the groundwater and so capable of migrating far from its origin or it could be a gas, able to vent upwards or laterally. The effects of the contamination include toxicity, for example with the polynuclear aromatic hydrocarbons that are carcinogenic; chemical attack on building materials, for example the degradation of concretes by sulphates and chlorides; and the ingress of potentially explosive or asphyxiant gases into buildings, for example landfill gas or carbon dioxide.

How much land is contaminated in the United Kingdom is uncertain, although an estimate of 100,000 to 200,000 ha has been made.[5] Experience in other countries suggests that only a relatively small proportion of potentially contaminated land presents an immediate threat to human health or to the environment. Within the generic category of chemical contamination there are two types of contamination that are often treated separately, asbestos and radioactivity, both of which are the subject of specific health and safety guidance.

---

5   *POST Contaminated Land*, Parliamentary Office of Science and Technology, London, 1993.

When they do occur, the problems caused if not the actual risk can become significant to the project, especially if no allowance has been made for the possibility of their presence. Gases can also be considered separately whether they are natural in origin or the result of human activity,[6] e.g. landfill gas; the risks they present are mainly of explosion and asphyxiation, the pathways are few, but even if the migration routes are far from obvious, the target to be protected, in effect people and a confined space, is amenable to identification and preventive design.

The consequences of chemical contamination to human health and safety are loss of life or limb, acute and chronic poisoning by inhalation, ingestion or absorption; and the environmental impact can be measured by the change in groundwater and surface water quality, damage to flora and fauna (ecotoxicity) and how the contaminants migrate to neighbouring land.

Assessment of risk

Contaminated land risks are often analysed using a source (or hazard) – pathway – target (or receptor) approach. Whether and how a contaminant presents a risk to a particular target or receptor (human health, building fabric, groundwater quality, etc.) depends on:

- contaminant concentrations
- the availability of the contaminant
- pathways to the target
- exposure time of the target to the contaminant or hazard presented by the contaminant
- sensitivity of the target to the contaminant or hazard presented by the contaminant.

---

[6] Hooker, P.J. & Bannon, M.P., *Methane: its occurrences and hazards in construction*, Report 130, CIRIA, London, 1993.

An assessment of the risks to human health and to the environment from the contaminant(s) or associated hazards can be given:

- as a qualitative statement which makes reference to the potential for harm
- by reference to, and comparison with, published guidelines[7]
- as calculated values using numerical models to simulate the processes whereby the contaminant impacts on specified targets or receptors
- by reference to statutory standards, for instance those for drinking water.[8]

If assessing risk is difficult, communicating the results often seems impossible. For any project there is likely to be a number of interested parties, each of whom usually has different views and perspectives on the implications for development and land use. These include the general public, local residents, the developer, local government and financiers. The development of a comprehensive risk management plan which takes into account not only the strict technical risk, but also the concerns of the various interested parties, is one option. Explaining clearly the reasons for taking particular actions and when those actions are especially visible or likely to cause temporary inconvenience, consulting beforehand, is vital if delay or simply bad publicity is to be avoided. These political, public and commercial pressures are heightened when contamination is the ground risk. In this context, note that in specific circumstances the US Environmental Protection Agency can award grants[9] for

---

[7]   Interdepartmental Committee on the Redevelopment of Contaminated Land (ICRCL), *Guidance on the assessment and redevelopment of contaminate land,* Guidance Note 59/83: 2nd Edition, July 1987.

[8]   EC Directive, 80/778/EEC

[9]   *The Superfund Program: Ten years of progress,* United States Environmental Protection Agency, EPA/540/8-91/003, June 1991.

technical assistance to help ensure that the local community is well informed about the condition of the site and the actions being taken and that they can be represented in discussions by an expert.

Management strategies

In contrast to the more usual types of ground risk that are amenable to deterministic or, less often, probabilistic methods of analysis, chemical contamination presents a different class of problem in that:

- remedial solutions are not yet proven in the long-term
- 'clean' is not an absolute value, nor a scientific term, but a subjective opinion, i.e. a political, policy decision
- many of the geochemical interactions are not fully understood
- the fate of chemical contaminants in the environment is often uncertain
- contaminants can migrate, within and off the site.

In order to deal with the risks of contamination, it is first necessary to resolve the question: 'does the site need treating and, if so, how best can this be achieved within any cost and time restrictions imposed on the project?'. If the site does need treating, which should be answered in the initial assessment of the risks, the various options for treatment have to be evaluated. Given the range of techniques available and being developed to treat soil and groundwater contamination,[10] a logical and disciplined approach to selection, as illustrated in

---

[10]   Arminshaw, R. *et al. Review of innovative contaminated soil clean-up process*, Warren Spring Laboratory, LR819, 1992.

Figure 1,[11] allows for objectives to be achieved while taking account of the various uncertainties and any constraints.

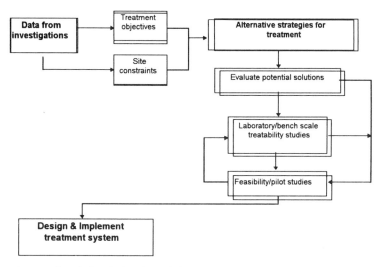

Figure 1   Steps in the selection of remedial treatment

Assessing treatment options does not necessarily imply a simple choice between a number of predefined methods, rather it invites a systems approach to the problem, for example soils washing with downstream aerobic digestion of organics and the use of chemical extraction of metals contamination.[12] Contamination-related objectives derived from the assessment of risks form the basis for developing remedial treatment plans and should specify:

- the contaminant(s) of concern
- exposure route(s) and target(s)

---

11   Johnson, S.T. & Laidler, D.W., *Contaminated Land, Business and the Environment,* Paper presented at IWEM Symposium, *Contaminated Land: Liability to Asset,* Birmingham, February 1994.

12   Bardos, R.P. & Harper, G., *The Integrated approach to cleaning contaminated soils,* Warren Spring Laboratory, Paper W91170, November 1991.

- an acceptable contaminant level, or range of levels, for each exposure route.

In practice there may also be engineering or management objectives to be met, related to physical works on the site or completions within defined periods, and site-specific constraints to be considered. Thus, the cost of the work, the practical difficulties of implementation, the need to reconcile often conflicting objectives, and legal or community-based objections to the use of a technically satisfactory solution, all have an impact on the selection process. Documenting evaluation and screening activities aids justification and the communication of ideas, results and decisions to all relevant parties.

The objective of all parties responsible for taking remedial action is to achieve a viable balance of safety, effectiveness and cost in the short term, and satisfy long-term requirements to minimise current and future liabilities. Options whereby the soil/groundwater is treated, either on or off the site, will become more attractive, particularly where there are specific physical constraints to excavation or where the nearest licensed site for disposal is some distance away. However well they are managed on behalf of immediate stakeholders, shareholders and investors, risks to the environment often motivate neighbouring landowners and local residents to object to schemes for treatment or to the transport of excavated material being taken off site through their community. Assessing at an early stage what degree of risk these parties, who generally lie outside business relationships of the project, might accept should form part of the overall assessment of risk and the management strategy adopted.

Viewing Ground Risks

Traditionally in geotechnical engineering, risk evaluation has used a deterministic approach, the calculation of a factor of safety against sliding or bearing failure for example. This type of analysis tends to hide the degree of uncertainty by use of 'blanket' factors of safety, even when separate (partial) factors are used for different components. While not often fully applied, Peck's[13] observational method to link design and construction to cope with the uncertainty of the ground combines analytical methods and monitoring. The principle is to modify the work in the light of what is being revealed about the ground and how it behaves. Observation has always been part of the practice of geotechnical engineering. In grouting, and lately in the New Austrian Tunnelling Method (NATM), the need to integrate the responsive feedback loops on site in the chosen contractual framework is often not fully appreciated.[14] Basically the contract has to allow changing the design as the work is built. With more sophisticated electronic measurement and control systems now available there is increasing scope for this way of working, as against the procedures involved in compensation grouting.

Still little used in geotechnical engineering, but gradually developing from other disciplines and the requirements of nuclear and defence clients, are various probabilistic analytical techniques. The potential of these methods will be further realised when geotechnical engineers are clearer about the types of uncertainty in the models they use and when there is greater understanding of how to assess the reliability of different geotechnical systems.[15] There are probably more uncertainties

---

13 Peck, R.B., *Advantages and limitations of the observational method in applied soil mechanics*, Ninth Rankine Lecture, *Geotechnique*, Vol **19**, No. 2, pp 177-87.

14 Powderham, A.J., *An overview of the observational method: development in cut-and-cover bored tunnelling projects*. [Accepted for publication in *Geotechnique* Symposium in Print, *The observational method in geotechnical engineering*.]

15 Chowdhury, R.N., *Evaluating risk*, Keynote Address to Conference on *Risk and reliability in ground engineering*, Institution of Civil Engineers, 11/12 November 1993, London.

and less understanding about the models used by environmental scientists in appraising the risks from ground contamination.

Type of damage	Who or what affected	Example ground condition
Technical (design or workman-ship)	Function Structure	Differential settlement Clay swelling/shrinkage
Contractual	Programme Budget Communication	Unforeseen obstructions Archaeological remains Founding depth uncertainty
Health	Workers Public Users/occupiers	Compressed air working Groundwater pollution Radon
Safety	Workers Public User/occupiers	Confined spaces Collapse of retaining wall Landfill gas
Property	Temporary works Permanent works Neighbouring property	Trench collapse Subsidence Loss of ground
Legal (prosecution or suit)	Company/organisation Individual	Lack of trench support Pollution of groundwater
Cost	Client Contractor Designer Tax payer Owner	Cost of changing method of working Extra support costs High PI premium Preventative and stabilisation works 'Escapes' to neighbouring land

Table 2 Types of damage and who or what could be put at risk

As geotechnical engineers we are considering technical systems, for example a foundation in which the ground and structure interact. But the risks are more than of technical inadequacy of a structure if they affect individual people, organisations and society. The types of damage that result from an accident or unplanned failure event[16] are summarised in Table 2 with examples of who or what might be affected, i.e. put at risk, and of the type of ground condition that might contribute to the hazard. The types of damage are quite general and apply to situations such as a fault of design, manufacture or installation of a steel beam or a concrete floor. Why, therefore, is the ground any different from any other source of risk? And why is contaminated land perceived as potentially even more risky?

Muir Wood[17] quoted by Blockley[18] highlighted three types of uncertainty about the geology – which here we will extend to the ground generally and so include artefacts. These are:

- unidentified features of the ground which may lead to behaviour different from the assumptions (incompleteness);
- identified features which might not be quantifiable or represented adequately in the risk model (system uncertainty); and
- failures of communication (human factors).

We shall examine each of these uncertainties in relation to the ground generally and to the possible presence of contaminants in the following section. It is to be remembered that these uncertainties are not separate issues: one uncertainty can compound others.

---

[16]   See Blockley *Op. cit.*
[17]   *Op. cit.*
[18]   *Op. cit.*

Incompleteness, System and Human Uncertainty

It is sobering to find out what proportion of the relevant ground volume is sampled even in a comprehensive site investigation for a construction project. Is the proportion 1/100, 1/1000 or 1/1,000,000? Even when a high proportion is sampled, 1/10,000 would be a very high proportion, a huge amount remains uninvestigated or to some extent unknown. Of course much can be inferred about the unsampled ground, and if the right people are asked to draw the inferences there may well be no problem from there being only a small sampling programme.

If the ground information is always incomplete, it only matters when someone has to find the reason, or an excuse, for a failure or damage (Table 2). It is sometimes easier to find an excuse out of a surfeit of information than to look for new information. Both are examples of incompleteness. The information may be available, but there are inadequate linkages between different characteristics, rock intact strength and fracturing for example, such that a harder rock with closer fractures may be easier to excavate and transport than a massively jointed but weaker rock.

Two further examples of incompleteness can be distinguished. One is when it is only by good fortune that a feature could be found, there being no reason to suspect its presence. There are very few true instances of this. Usually there would be grounds for suspicion, the crucial step before recognising risk, as, for example, a general awareness of there being solution features or bell pits in certain chalk areas. A more heinous type of incompleteness is when there is information available for the asking, like an old map or air photo that shows a new-filled canal or demolished building. This is also a failure of communication and of the education of the person doing the desk study or who should have been commissioning a proper desk study.

System uncertainties are easier to explain by considering working operations, but they apply to geotechnical design as well. The typical particle size distribution ranges of an Upper Mercian Mudstone (Keuper Marl) might well have been identified, but its susceptibility as an earthworks material to moisture changes remains a difficulty. Very little rain on opened-up material in a borrow area or when being spread at fill placement can rapidly reduce potentially good earthfill to a slurry. It is particularly where the chosen method of working proves unsatisfactory and has to be modified or completely changed that the greatest blame is put upon the ground or on the information provided about the ground. CIRIA Report 79[19] tried to address more equitable methods of sharing risk for tunnelling, where the choice of method is critical. At a different range of initial cost but with the potential for expensive delays is the choice of method of piling, or dewatering, or groundwater exclusion.

When dealing with a potentially contaminated site the same general points apply.

- Identification of past use provides very good clues to the possibility of particular forms of hazard being present.

- There will remain the practicability problem of a close enough search to identify the locations and types of all the contamination.

- System uncertainties will be of various forms even though the presence of specific contamination is known. Thus how a contaminant may now or in the future be distributed through a ground mass or groundwater body might not be predictable. The

---

[19] *Tunnelling, improved contract practices*, Report 79, CIRIA, London, 1978.

selection of a remedial treatment method to deal with one form of contamination might be impeded or inhibited by the presence of other materials in the ground or at least need costly modifications. The more sophisticated the technology of the method, often the more susceptible it is to conditions differing from those it was designed to treat. The comparison can be made with tunnelling methods. Tunnel boring machines, while now increasingly capable of coping with a range of ground conditions, do not yet have the inherent flexibility of labour-based, drill-and-blast or compressed air shield works. A large area of system uncertainty is embodied in the question, 'how clean is clean?'. Even where there are trigger levels (which seem analogous to deterministic safety factors), what do they mean? how are concentrations to be measured? what degree of protection to the environment do they give?

- Again, of course, there are the uncertainties of human factors and inadequate communication. Part of this is the problem wrapped up in the phrase 'perceived risk'. The 'perception' of risk or of the possibility of an accident is the start of safety, as with the suspicion that in the chalk below the shallow overburden there may be solution features. What tends to be meant by 'perceived risk' is a misperception either that the hazardous condition presents an insuperable problem or that the probability of loss is particularly high. Emotive words are used, contamination is one, and judgements are subjective. It is not always appreciated that as well as artificial materials there are naturally occurring substances that 'contaminate' the ground. Methane as marsh gas or mine gas is natural, but artificial when in landfill gas. Radon, heavy metals, trace elements, sulphates and carbon dioxide, sulphides and, even, oil seeps can be present naturally and it is not always readily apparent as to how much of a measured concentration of a contaminant is natural or artificial.[20]

- Emphasised in every report that has sought improvements in construction practices and particularly in relation to site and ground investigation is the need to choose an appropriate system of procurement, in other words to get the right people to do the right things at the right time for the right reasons (only one of which is the right reimbursement!).

Perhaps too much of a conjecture, but with more than a seed of truth in them are the statements:

- most of the geotechnical engineering (in UK) is done by non-geotechnical engineers;
- most of the time of good geotechnical engineers is spent in the resolution of claims, disputes and litigation consequent upon there not having been sufficient time or thought given to a project by a good geotechnical engineer.

Contrarily, there are more geotechnical engineers now than ever before, more efforts to stipulate who can be considered a geotechnical engineer, more efforts put into quality management, and more knowledge of ground materials, their properties and behaviour. The same remarks can be made, with even more truth, about the environmental engineering of the ground. The key differences, however, are:

---

[20] *Review of natural contamination in Great Britian.* Current project in Geological and Minerals, Planning Research Programme: Land instability and safety, Minerals Division, Department of the Environment.

- there are very few remedial treatment specialists;
- ground contamination problems usually require multi-disciplinary teamwork; and
- we know less about the long-term consequences and therefore probably do not appreciate all the problems of contamination.

The August 1994 issue of *Civil Engineering* highlights the topical arguments that follow from the above differences. Wood and Griffith[21] debate the balance between cost and caution in redeveloping contaminated sites.

Engineering and Expert Judgment

In a most enlightening paper, Skipp and Woo[22] explain for geotechnical engineers the distinction between engineering judgement, all too often the catch-all cover-up of a whole group of uncertainties, and expert judgement, which is a formal and mathematically rigorous elicitation process to enumerate subjective probabilities. As these authors point out.

> *"... Where an engineering judgement is unwittingly offered in place of an expert judgement, or vice versa, problems can be expected. One area of potential importance is litigation. As [probability] methods become accepted as a means for weighing evidence in courts of law, there will be a need to clarify the types of*

---

21   Wood, A.A. & Griffiths, C.M., *Debate: contaminated sites are being over-engineered*, Proceedings Institution of Civil Engineers, (1994) 102 Civil Engineering, No.3, August, pp.97-105.

22   Skipp, B.O. & Woo, G., *A question of judgment: expert or engineering?*, In *Risk and reliability in ground engineering*, pp.29-39, Thomas Telford, London, 1993.

judgement offered by professional engineers in this context." [23]

Table 3 shows the attributes of the two types of judgement. What becomes clear from an example given by Skipp and Woo is that just one geotechnical aspect of a project, for example an assessment of slope stability, involves numerous technical judgements, both engineering and expert, and commercial, management and regulatory (compliance) judgements.

Engineering judgement	Expert judgement
Not constrained by formal logic	Constrained by formal logic
Cannot be easily calibrated	Can be calibrated
May incorporate conservatisms	Must be true opinion
Tends to heuristic and holistic	Focuses on parameters

Table 3   Attributes of engineering judgement and expert judgement

Client Tools for Managing Risk

The principal risk to clients is financial, through:

- time delays;
- increased cost of materials, etc;
- bank charges;
- legal liabilities; and
- shareholder responsibilities.

Ultimately it is the client who dictates the overall approach within any project to the sharing and apportionment of risks in ground engineering. In dealing with most ground risks and

---

[23]   *Ibid p.38*

particularly with chemical contamination a serious constraint on client actions and reactions to problems is confidence. This manifests itself in the assessment and subsequent communication of the nature of the problem and in the efficacy of the technical solution. Invariably these concerns have a marked effect on the final value of the land and to future liabilities. Evaluating the risk for the client is about liabilities, both actual and potential to themselves and to other stakeholders, shareholders and investors in the project or the company. On a particular site the impact of a geohazard such as a slope failure will be assessed with regard to:

- contract conditions entered into by the consultant and contractor and, by implication, by any specialist subcontractor;
- existence of appropriate warranties and collateral warranties;
- bonds and other similar assurances;
- insurance;
- the common law position; and
- the regulatory compliance position.

For each of the interested parties represented by or contracted to the client the consequences of the eventuality of the hazard depend on the overall business structure and specifically assigned roles and responsibilities.

Contamination of ground and groundwater

Chemical contamination risks present particular problems for clients. When contamination is suspected or found then it may be tempting to abandon the project and do nothing. However the do-nothing option could also incur current legal liability, the consequences of which may include the payment of a fine or a compensatory payment, an obligation to pay for remedial work carried out by a regulatory body or a legal requirement to take remedial action. For lenders the three main areas of concern in relation to contamination are fluctuations in the value of the land or property, the ability of the borrower to repay loans, and any primary liability for loss and remedial action. The last concern is a particular risk , one where the lender has to enforce security and take possession of a property along with the potential liabilities related to the contamination. A similar situation could also exist where the lender intervenes in the management of a company where there is contamination of the land, almost irrespective of the nature of the intervention.

In the management of all risks and particularly those of chemical contamination, insurance has a potentially key role to play. There are perhaps four main types of specialist policy either available or proposed:

- third party legal liability;
- public liability for contractors;
- underground storage tank liability; and
- site-specific insurance-backed warranties.

Almost without exception these policies are written on a claims-made basis with no retroactive date provision and until recently could not be extended beyond five years in the first instance. Most policies are renewable annually. At present this specialist market is perceived as relatively expensive by clients and consultants for the cover available. However as insurance for pollution and environmental impairment becomes restricted on renewal of existing policies, these specialist policies are likely to become more attractive, in the absence of alternative ways to limit liability.

Professional indemnity

The key attributes of professional indemnity (PI) insurance are:

- company-wide cover;
- annual renewal;
- written on a claims-made basis;
- insurance is against proven negligence; and
- usually subject to an excess.

Traditionally, the indemnity policy has provided the consultant and other professional advisors with the necessary long-stop to meet claims of professional negligence, including errors or omissions resulting from the professional services specified in the policy. Making a claim on any PI policy, over and above the excess, requires negligence to be proven. This is not a ready remedy either for the client, who will be paying for any extra work in advance of compensation, or for the consultant/contractor in terms of the costs involved in defending the claim in court. An over-reliance by clients on this particular form of insurance would appear to be misplaced, not only in the time it can take to reach a conclusion following a claim, but also the difficulty in many instances of proving negligence. Unfortunately even when the court does not find against the insured, there can be high costs, directly in terms of loss of fee-earning time of senior staff, legal costs if not borne by the insurance company or rising premiums and the prospect of future exclusion clauses.

There are moves to limit PI cover to exclude all pollution (contamination) and related claims for environmental damage as individual policies come up for annual renewal. Other approaches, therefore, need to be considered so that the right level of protection is available to the right people for the right events at the right price. Tyrell[24] in a recent article for the

---

[24] Tyrrell, A.P., *Consultant's pollution liability: A health warning,* Association of Geotechnical Specialists Newsletter, Issue No. 15, March 1994.

Association of Geotechnical Specialists suggests two main options:

1. Specific exclusion of liability for any losses incurred by the client in respect of pollution or contamination and/or arising out of or in connection with pollution or contamination.

2. Limitation of the total liability of the consultants for any claim arising out of or in connection with pollution or contamination to the least of the following:

    *   an amount which takes into account factors such as the consultant's fee and the value of the project
    *   the cover provided by the professional indemnity insurance policy
    *   remediation costs alone, consequential losses being excluded.

It is important irrespective of the cover obtained, for all parties to know the extent of each other's liabilities to ensure that the project is adequately covered. A major concern is over past work carried out but not covered either directly by today's policies or through any associated warranties, whereby claims made will be under the new policy and not the old one.

Procurement of Ground and Environmental Services

Investigation and consultancy

Last year the Site Investigation Steering Group, which embodies representatives of 21 organisations under the chairmanship of Professor G.S.Littlejohn, published four documents[25] to help improve the quality of site investigation and to obtain better value for money for the construction industry and its clients. The organisations represented include government departments, learned societies, professional institutions and trade association, covering the width of the construction community not just the geotechnical specialists, although inevitably they have provided much of the input. This is the most recent of many attempts to create better awareness of the need for thorough site investigation. It should be noted that the fourth document provides guidelines for investigative drilling of contaminated sites.

In the mid 1980s, a study by Uff and Clayton,[26] jointly commissioned by CIRIA and BRE with PSA support, of ways to improve ground investigation concluded that the underlying reason for inadequate investigation is inappropriate procurement. Widely and sometimes hotly debated, their proposals of alternative, but distinctly different, procurement systems stemmed from a recognition that, too often, the procurement procedures, the people involved, and the investigation practices are too remote from the proof of the investigation: the performance of what is built. Both of the proposed systems bring investigation closer to its consequences. The established system, with conventional forms of contract and the separate employment of a contractor for the physical work, depends upon proper definition of the role of the geotechnical adviser.

The second proposed system was for a single contract for geotechnical expertise, together with physical work, testing and reporting. A new contract form and procedures to be followed were put forward in the report, a feature of it being that the

25   *Site investigation in construction*, Site Investigation Steering Group, Thomas Telford, London, 1993

26   Uff, J.F. & Clayton, C.R.I., *Recommendations for the procurement of ground investigation*, Special Publication 45, CIRIA, London, 1986.

specialist who designs, carries out and reports the investigation is required to give a specific warranty. In addition, the standard form of dispute clause was modified to allow any claim against the specialist to be brought concurrently with any dispute under the main works contract which relates to the investigation.

It seems, however, that this second procurement system, akin to 'design-and-build' for ground investigation, was not tested in practice. Although that might seem disappointing, the debate in the geotechnical community that the project raised and the subsequent report *Role and responsibility in site investigation*[27] by the same authors, together with the introduction of quality assurance requirements, led to three developments:

- the publication by the British Geotechnical Society of a register of geotechnical specialists;
- the formation of the Association of Geotechnical Specialists, an organisation which although a trade association, cuts across traditional boundaries of consultant, contractor, specialist subcontractor, engineer, geologist, etc; and
- the tempering of initial proposals for quality assurance systems to more practicable and viable procedures.

Yet none of these initiatives has had much effect on the bulk of site investigation works. Piling specialists today complain that the ground investigation reports are no better, and many would say, much worse than they were ten years ago. Hence the support given to the Site Investigation Steering Group.[28]

Over the last 30 years, site investigation practitioners have put considerable effort to improving their work and to encouraging

---

[27]   Uff, J.F. & Clayton, C.R.I., Special Publication 73, CIRIA, London, 1991.
[28]   *Op. cit*

others to give more thought to the requirements of good investigation. It seems to little avail. The awareness now of the potential for sites being contaminated and the need to extend the investigation methods to cope with the numerous contaminants and their different forms has changed the focus of self-help and brought in people from other disciplines with other skills. There is now, for example, the Association of Environmental Consultants whose membership has some overlap with that of the Association of Geotechnical Specialists, but in the main is different, and which would claim a much wider remit than geotechnical engineering.

In some ways this has created healthy competition, but in others a damaging rivalry with the potential for conflicting advice to clients. The enlightened client will have staff or existing advisors to help appoint a suitable team. But at a time of free market competition, when there are many more firms and individuals offering contaminated land expertise than there have been completed remedial treatment projects, hard questions have to be asked before commissioning new advisers. It seems all too likely that the history of site investigation will be repeated in environmental investigation.

Perhaps the clearest lesson that has come out of the site investigation debates is the value of thorough desk study to find out as much about the natural and human history of the site as possible before designing the exploratory work. Thorough desk study is even more important when there is a suspicion of contamination from past use of the site or its environs.

Specialist contractors

Many of the risks of the ground first arise when specialist ground engineering operations in effect open up the ground, in foundation construction, grouting, dewatering, etc. It is because of the need to adjust treatments, whether of grout injection or wellpoint spacing, as the work proceeds to achieve the necessary performance that the techniques themselves are considered as ground risks. Very often these works are subcontracts and so a delay in the ground engineering work has an immediate consequence for the main works. The link is direct and obvious. Yet it is particularly those types of ground treatment that rely upon close observation of ground response and so take the opportunity to modify the technique for greater efficiency, in effect altering the design during construction which are seen as high risk.

If traditional prejudices are put to one side as they will have to be if partnering arrangements are to succeed and when design-and-construct becomes more common, it could well be that the Observational Method, in the sense meant by Peck[29] and illustrated by Powderham,[30] will be increasingly used. Properly set up and coupled with value engineering it provides a rational way of coping with many of the uncertainties of the ground and of contamination. If, however, remedial treatment is handled in the same way as is so much of the ground treatment, i.e. as a domestic subcontract, it will prove to be yet another cause of claims and delays.

A Framework for Assessment

There are so many possible ground and contamination hazards that it is convenient to represent them and how they are managed by a simple construct. The hazards, as we have seen, are in the ground, in the contamination and in their mutual interactions and with the ground engineering of the project. Risk interactions as pointed are two-way processes, thus by its intrusion contamination changes the nature of the ground but, in

---

[29]   *Op. cit.*
[30]   *Op. cit*

turn, the ground can change the nature of the contamination, by dispersal in groundwater for example. Just as the engineering options can alter the ground, making it less permeable by grouting, so the remedial treatment might remove or contain contamination. But it is the nature of the ground and the presence of contamination which determines what treatment works can be used for the chosen use of the site.

Interaction matrix

Constructing an interaction matrix[31] is a good way to summarise these considerations of ground risk management (Figure 2). The main diagonal boxes are the primary factors, (A) the ground (in this case including the groundwater, although it properly should have its separate identity), (B) the contamination, (C) the engineering and (D) risk management as the primary purpose. Separate interactions are shown as between primary factor (A) on primary factors (B), (C) and (D) and the complementary (and different) interactions of primary factors (B), (C) and (D) respectively on primary factor (A). Thus A → B ≠ B → A. With some effort but to telling effect in achieving clearer thinking about ground risks, the interaction matrix can be expanded by other important factors, e.g. groundwater, time or even different types of engineering works, such as those to treat the contamination and those to improve poor ground. It would also be worthwhile to examine the implications for different procurement systems.

---

[31] Hudson, J.A., *Rock mechanics principles in engineering practice*, CIRIA Butterworth, London, 1989.

Risk management framework[32]

In Figure 2 the last column shows how the risk management objectives for technical and environmental performance set by the ground and the contamination add to the commercial and legal objectives that the engineering project has to achieve. The bottom row of the figure represents the influence of risk management choices first in achieving the acceptable balance between cost and caution and, secondly, in setting the standards to be reached by the ground engineering and remedial treatment works.

---

[32] For a detailed explanation of interaction matrices, see Hudson, J.A. *Rock Engineering Systems: Theory and Practice*, Ellis Horwood, London, 1992

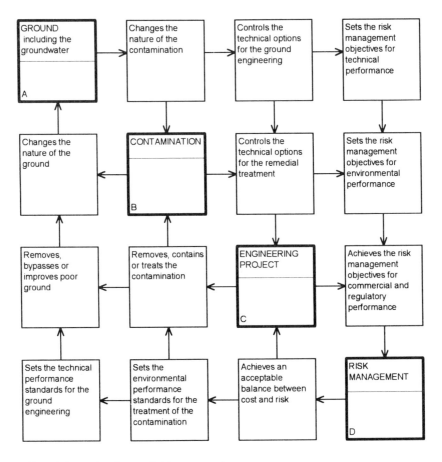

Figure 2 Interactions in ground risk management

Information and guidance framework

Figure 2 also illustrates CIRIA's role in ground and contaminated land engineering as a provider of research information and best practice guidance to the construction industry. CIRIA is able to perform this role by the support of its members, government departments, particularly Department of the Environment, and other industry sponsors. Objectives of better design, more efficient construction, higher productivity, value engineering savings, minimal environmental impact, etc.

underlie all CIRIA research projects. In one way or other these are included in the boxes of the right hand column of Figure 2. The guidance given in CIRIA reports aims to help construction practitioners and, hence their clients, achieve appropriate technical, environmental, commercial and regulatory objectives. When, and often this is the justification for the project, formal codes and standards are not yet available, CIRIA research projects help to establish what is accepted as good practice, and so to set the technical and environmental performance standards without which a rational judgement of the balance between cost and risk is not possible. The guidance is there for those less experienced, *i.e.* those who do most of the geotechnical engineering The establishment of good practice not only provides a benchmark for the less capable to strive for, it can be the stepping-off point to engineering advances. In themselves these are justifications of CIRIA's research programmes, even if geotechnical engineering continues to give income to lawyers.

11. Ground Uncertainty Effects on Project Finance

Ian Whyte[*]

Synopsis

This paper analyses the economic consequences of uncertainty regarding ground conditions. The manner in which geotechnical investigation can lead to an increased level of certainty is described. Management of the risks arising from ground conditions is examined in the context of a tunnelling case study. Conclusions are made as to the relationship between site investigation and economic risk.

Introduction

Construction projects such as roads, bridges, buildings, industrial development and other civil engineering works are all built on or in the ground. Site investigations are made to forecast the ground conditions for planning, design and construction purposes. Even with a full and adequate investigation it is not possible to be certain about the nature, distribution and physical properties of the ground and groundwater. An element of uncertainty is inherent to all construction works, the skill of the engineer is to reduce the level of

[*] The author acknowledges the work of Dr. W.S. Peacock and Dr. N.M.H. Alhalaby. Support for the research was provided by SERC and thanks are given to those individuals and organisations who provided information for the case studies.

uncertainty to acceptable limits. Unfortunately, some investigations are less than adequate and projects become exposed to ground related problems which may not have been foreseen. The economic consequences of ground difficulties can be significant in terms of both cost and delay. Research has shown that the level of financial risk with a project reduces significantly with properly planned and adequate investigation. A proper engineered approach has direct financial benefit due to the reduction in uncertainty and improved reliability in cost forecasting.

The greatest uncertainty to a project is present during its early stage of development, risks have to be identified, analysed and managed. Geotechnical works are particularly high risk activities and are most uncertain at the early planning stage of a project. As information is obtained during project development then problems become more defined and uncertainty levels decrease, the degree of risk reduction is then a function of the adequacy of the works. Methods of site investigation have evolved to a high level of technical sophistication. Such methods are most frequently applied to prestigious and large-scale works, less so to more routine construction. Practices that are traditional and which originated many years ago are common and are not questioned critically. Progress to improved procedures can be hindered by a low level of appreciation of the financial consequences leading to poor management, low levels of resource, faulty communications, ill-defined responsibilities and methods of procurement and contracts which are simplistic.

The development of site investigation practice in the UK

It is necessary to appreciate how the site investigation industry has developed in order to understand the reasons for concern. Such knowledge is required also if progress is to be made in a sequential manner rather than by a revolution. Revolutions do have a tendency to throw up new problems which do not necessarily solve the defects in systems. Prior to the 20th century, investigations were undertaken for major projects with one of the earliest referenced[1] being to borings for a canal in China around 1066. Engineers bored holes, probed the ground and designed largely on the basis of art and precedence rather than by science.[2] Practices for drilling, sampling and probing had by the 17th and 18th centuries reached levels not too dissimilar to present day techniques.[3] At the start of the 20th century, major ground difficulties were encountered with the construction of large projects such as the Kiev canal in Germany, railways in Sweden and Panama canal. Terzaghi developed the science of soil mechanics in the 1920's and gradually a scientific and engineered approach evolved, particularly in America and Sweden.

In the UK, current site investigation practice came out of the failure of Chingford Dam in 1937 where Terzaghi acted as technical adviser and commissioned boring, sampling and laboratory testing. The contractor spotted a business opportunity and soon formed a specialist company to market the services. The company was called Soil Mechanics Limited and is still a market leader. Other contractors in the post-war period followed the example set. Routine investigatory work evolved as a contractual service commissioned by competitive tender. Conditions of contract were not formalised and rarely used, it was not until 1983 that special conditions were published and in 1989 a general specification and bill of quantities was produced. Practices in the industry followed those formulated in the early years and, in essence, remain unchanged to the present day. Many of the method specifications, for example, those for undisturbed soil sampling, pre-dated research

[1] Needham J., *Science and Civilisation in China*, Vol.4, part III, p. 333, Cambridge Univ. Press, 1971
[2] Parnell H. *A Treatise on Roads*, Longman, London, 1833; Skempton A.W. *Significance of Terzaghi's Concept of Effective Stress*. From *Theory to Practice in Soil Mechanics*, Wiley and Sons, New York, 1960
[3] Jensen M. *Civil Engineering around 1700*, Danish Technical Press, Copenhagen, 1969

into quality and thus present operations are not always reliable and can even mislead.

In the 1980's, the price competition produced severe financial problems and several firms closed down. Others sold their drilling operations and the operatives offered sub-contract services for site works. The industry was fragmenting with increasing uncertainty over responsibilities. Clients became unsatisfied with the product.[4] A BRE/CIRIA report[5] in 1986 confirmed the industry suffered from work of indifferent quality, low price levels, insufficient resources and lack of investment in training. A working party in 1987[6] developed principles for good practice and recommended a change in attitude from work execution at minimal cost to fitness for purpose and reliability of outcome. A survey reported in 1990[7] showed the following:

- Many forms of procedure exist for the procurement of investigations, some of which are not suited for obtaining reliable information;
- Managers for site investigations (ie. those who control the resources) are normally appointed to suit the construction project and generally have little to no geotechnical expertise. Qualified and experienced geotechnical managers are the exception rather than the rule;
- The financial risks associated with site investigations are not fully understood by clients;

---

[4] Rys, L.J., Wood I.R. *A Question of Priority in Product before Procurement: Site Inv. in Practice: Assessing BS5930*, Proc. 20th Reg. Mtg Engng Gp. Geol. Soc., Univ. of Surrey 1984, Geological Society, 1986; Peacock W.S. & Whyte I.L. *site investigation Practice*, Mun. Engr. 1988, Oct 235-245

[5] Uff J.F. & Clayton C.R.I. *Recommendations for the Procurement of Ground Investigations*, SP45 CIRIA, London, 1986

[6] Buttfield A. *Future Site Investigations - Quality and Value for Money*, Civil Engng. May 1987, 63-65

[7] Peacock W.S. *Site Investigation Procedures and Risk Analysis*, UMIST PhD thesis, 1990

- In the 1980's, there was an increase in the use of competitive tendering and rigid forms of specifications.

In 1991, the Institution of Civil Engineers produced an authoritive report[8] on inadequate site investigations. After this, the following development occurred:

- The publication by the British Geotechnical Society of a register of geotechnical engineers.
- The formation of an Association of Geotechnical Specialists and the British Drilling Association as trade organisations. These promote standards of good practice and quality assurance schemes, for example the register of accredited drillers.
- The formation of a site investigation steering group, under the aegis of ICE, with representation by clients, consultants, contractors, specialists. This group produced four agreed reports[9] in 1994. Report 1 is aimed at clients and non-specialists and explains why the ground can be a hazard. This document is produced at a low price to encourage maximum circulation and can be given to clients at insignificant cost.

It is evident that whilst the principles of good practice are now being clearly stated and published, there is still little change in most of the industry. Recent procurement changes may even cause a deterioration of standard, for example competitive tendering of professional services and transfer of ground risk from client to contractor under design and build systems. There is a real problem in the conflict between the demand from auditors and financiers for 'value for money', 'open competition' and professional standards,

---

[8] Inst. Civ. Eng. *Inadequate site investigations*, Thomas Telford Ltd., 1991 26 *et seq*

[9] Inst. Civ. Eng. Site Investigation *in Construction: 4 reports*, Site Inv. Steering Gp., Thomas Telford Ltd., London 1994

required levels of management and technology for good standard investigations. The lesson has to be made that low cost does not equate with low risk. Rather low cost in ground management produces high uncertainty and exposure to unacceptable levels of cost overrun and delay. Attempts are now being made to quantify these issues.

Costs and delays resulting from ground uncertainty

Ground investigation is a low cost construction activity. Data accumulated over many years suggests that the costs are typically in a range of 0.2% to 0.5% of contract value and rarely exceed 1%. The cost consequences of 'unforeseen' ground events can often exceed 10% of contract value and in some recorded cases the additional costs have exceeded 100%. Not all risk events offer losses. This is exemplified by such events as a reduction in the rate of interest on a loan, or ground conditions being better than anticipated. Most hazards, however, have risks that produce an adverse effect and can result in instability, delay and losses The ground is known to present potentially high hazards and to be subject to uncertainty. Minimising the risk requires investment in professional costs and time, often at the earliest stages of a project when uncertainty is at its greatest level and key decisions have yet to be taken. It has to be emphasised that time as a resource for investigation is equal to cost. A failure to invest in an adequate knowledge of the ground and site history (the 'desk' study) can have significant effects on project economics and timing. Evidence published in the last 10 years or so support this. Examples which support this contention follow.

Tunnel

An American report[10] on 87 projects showed a direct correlation between the accuracy of a cost estimate (in relation to final out-turn costs) and the level of site investigation (site investigation) It was recommended that site investigation expenditure be increased to at least 3% of the estimated project cost. In the survey the average cost of investigations for tunnels was 0.44%. About 60% of the claims were for large amounts. Analysis showed that both the level and number of claims related inversely to the amount of investigation. The more known about the ground in advance of construction reduced the level of uncertainty leading to better planning and forecasting and a reduction in the number and degree of severity of the claims.

Guidelines for avoiding and resolving disputes in underground construction were reported in 1989[11] and were extended in 1991[12] to other works such as roads, bridges, buildings, etc. These procedures aimed at methods for developing co-operation on projects through an equitable risk sharing philosophy between client and contractor. Three main elements are involved:

a) A dispute review board

This consists of 3 members experienced in the work. They visit the jobs during construction and receive monthly progress reports. Should a dispute develop they can take immediate and informed action to obtain a quick resolution before attitudes become

---

[10] National Research Council. *Geotechnical site investigations for Underground Projects*, Vols. 1,2, Nat. Ac. Press, Washington D.C., 1984

[11] Technical Committee on Contracting Practices - *Underground Technology Research Council. Avoiding and Resolving Disputes in Underground Construction*, ASCE, New York, 1989

[12] Technical Committee on Contracting Practices - *Underground Technology Research Council. Avoiding and Resolving Disputes during Construction*, ASCE, New York, 1991

adversarial. The use of a dispute review board does not represent any saving on the works, the real savings come in the elimination of lengthy periods (often years) to resolve disputes. The cost of a DRB meeting is reported to be about $5000.

b) Escrow bid documents

These consist of the quantity take-offs, calculations, quotes, consultants reports, notes and other information used by a contractor to arrive at a price. The client can only gain access to the documents in the presence of a representative of the contractor. The usual practice is to use the documents during negotiations for equitable adjustments to contract price. Experience shows there to be no undue cost or inconvenience in preparing the documents for escrow storage.

c) Geotechnical design summary report (GDSR)

This report states in unequivocal terms the client's understanding of anticipated ground conditions and impact on design and construction. It forms the geotechnical baseline for the contract and emphasises openness and candour throughout the contract process. The use of a geotechnical design summary report has led to uniform bid prices with less exposure to claims involving interpretation of ground data.

Clients have expressed concern over the use of a GDSR. The key is the recognition that the document enables the client to appreciate how ground conditions influence his management of risk. A conservative baseline reduces the risk of unexpected cost increase but increases the contract bid prices. The client, under advice, has control over the trade-offs between risk and price. For the engineer, the GDSR demands a thorough geotechnical assessment of design and construction. It produces better practice than when the contract emphasis is on factual data alone. Experience with the above tripartite system is that the number of differing site conditions claims, and subsequent amount of financial award, are decreased significantly. It is reported that the practices have saved millions of dollars in the avoidance of claims and litigation. By 1991, the procedure had been used on over 100 contracts of over $6 billion value with very few disputes proceeding to arbitration or court.

Highways

In 1992, the National Audit Office report to the House of Commons[13] that the cost of highway contracts in 1990-91 increased by 28% over tender values. Nearly 50% of this increase was attributable to geotechnical conditions: an estimate of the geotechnical cost alone for that year was £100 million. These costs were considered to be unacceptable and more recent evidence on motorway widening projects suggests no improvement in ground cost escalation. The DoT commissioned a survey of site investigation costs on highway contracts. In addition, the NAO recommended that the ground risks not be assumed by the DoT but transferred to the contractor through the medium of design and build. It was judged that this could result in a bid cost escalation of 15%, however, these costs could be justifiable on the basis that they were subject to competitive market forces. Recent announcements in the press have shown the policy of design and build (with no ground risk to the client) to become more the norm for highway works with DoT and Highways Agency. It is too early at this stage to tell whether or not the philosophy is likely to succeed and give value for money. The principle of the client avoiding the ground risk by competitive contract, however, is controversial and at odds with management principles of fair risk allocation[14]

---

[13] National Audit Office Department of Transport. *Contracting for Roads*, HMSO, Dec. 1992, 39pp

[14] Perry J.G. & Hoare D.J. *Contracts of the Future: Risks and Rewards*. Future Directions in Construction Law, Proceedings of the 5th Annual Conference, Centre of Construction Law and Management, King's College, London 1992

Buildings

NEDO reports in the 1980's[15] showed that up to 50% of commercial and industrial developments experienced delay due to ground problems; and of those developments on 'brownfield' sites all experienced delay due to ground conditions. Furthermore, the findings of a report into serious delays[16] in these projects indicated that they were virtually always associated with the ground. Cost estimates for these ground problems were not reported. A large supermarket development could, however, expect a turnover in the range of £300K to £400K per week and thus a 10 week delay can result in a revenue loss of £3M to £4M, plus additional loan and finance charges. Such costs far exceed (by many factors) any investigation costs.

In the UK, the present situation with regard to procurement, procedure and contract is one of change. *Constructing the Team* illustrates that the construction industry has been greatly affected by recession, and that it is not easy to create teamwork and co-operation whilst struggling to avoid losses. Traditional contracts such as the ICE 5th and 6th editions tend to encourage adversarial situations as one side tries to score a claim against the other. New alternative are under review, such as the New Engineering Contract, design and build, BOT and BOOT schemes. The NEC may become a standard through parliamentary legislation. There is an increasing emphasis on risk strategies, definition of responsibilities, change, evaluation of quality (whole life issues) rather than price, effective planning and teamwork. For such measures to succeed an essential component to consider is the ground hazard. Management has to consider the effectiveness on other risk issues, particularly finance.

---

[15] NEDO *Faster Building for Industry*, National Economic Development Office London, 1983; *Faster Building for Commerce*, Nat. Ec. Dev. Off. London, 1988

[16] Taken to be delays in excess of 10 weeks

Managing the Risk

The perception of risk varies according to viewpoint: one man's loss may be another's gain. Engineers and construction professionals, designers and contractors, are trained to high degrees of proficiency in technology and tend to view hazards and risk from that viewpoint. Developers, and their banking and financial advisers, manage assets and markets and view events in terms of economic power and the market. It is evident that there is a problem in relation to ground risks. Insufficient resources of time and money produce inadequate investigations which have both technical and financial consequences. Much has been written as to what the problem is and engineers have sought solutions through procedural change and forms of contract. Jardine and Johnson state[17] that none of these initiatives has had much effect on the bulk of site investigation works. Piling specialists report that ground investigation reports are no better than ten years ago when standards were abysmally low.

Over the past 30 years, geotechnical engineers have put considerable effort into how to improve their work. It seems to little avail. At a time of free market competition for construction services with clients desiring to avoid the ground risks through contract conditions, warranties and insurance, there is no immediate hope for a solution to ground problems. One thing is, however, certain, the answer will require additional resources. Even though the cost implication is small, site investigation costs would increase to, of the order of, 1% or 2% of the tender value, such resources will not be forthcoming unless justified. Engineers have, therefore, to convince clients and their funders that there is a worthwhile return on the additional investment in ground studies. Only then can competent managers produce the goods for specialists to display their skills and knowledge. For this to happen, engineers must communicate their desires using language understandable to bankers and financiers.

---

[17] Jardine F. & Johnson S. *Risks in Ground Engineering: A framework for Assessment*, Paper No *10 supra*

Wood[18] and Scriven[19] have reviewed the bankers' view on project security and finance. The main concerns of financing institutions are:

- *Capital overruns.* This is the most common difficulty and mainly arises from inadequate preliminary engineering and poor estimating procedures. It is not due to inflation.

- *Completion delays.* Almost as common as capital overruns is the inability to complete the project on time.

- *Cash flow variations.* Caused by operating costs, variable productivity, currency fluctuations, equipment obsolescence.

- *Market difficulties.* Volatility in markets, imposed price controls, demand problems.

- *Resource risks.* Adequacy of reserves of oil, gas, minerals. Quality, quantity and structural mistakes.

- *Political actions.* Taxes, royalties, import duties, control, expropriation.

- *Project Inefficiency.* Poor project planning and management, poor management of completed facility.

---

[18] Wood P. *The Banking view of Project Power*. In *Future Directions in Construction Law*, Proceedings of the 5th Annual Conference, Centre of Construction Law and Management, King's College, London, 1992

[19] Scriven J. *A Banking Perspective on Construction Projects*. In *Future Developments in Construction Law*. Proceedings of the 5th Annual Conference, Centre for Construction Law and Management., King's College, London, 1992.

- *Environmental risks.* Pollution, contamination, migration of pollutants, liabilities. Inadequate environmental audit.

- *Casualties.* Accidents, incidents, natural processes which can by subject to insurance. Is cover adequate?

- *Insolvency.* Closure of contractors, sponsors, suppliers, purchasers, insurers, bankers.

The above list is not complete and other risk items can be added. Geotechnical engineering is a technology and would not necessarily be a risk choice by a banker or financier. The available evidence suggest, however, that ground risks pose the greatest uncertainty to the construction process and frequently influence activities on the critical path. Ground risks , therefore, are a significant feature of capital overruns and completion delays. When banks assess project proposals they do undertake risk analysis. Risks have to be clearly taken by individual parties to the project; risks have to minimised or avoided and the party that bears the risk has to have the resources / collateral to do so. The principles are :

- Risks attributable to investment issues, for example; political and general business risks. These can be covered by an increased discount rate and may not be shown in the cash flows. The risks are taken primarily by the banker or investor.
- Insurable risks, for example; technical risks connected with project construction and operation. These can be covered in the cash flow by a premium payable to cover such risks. The premium can be for insurance, alternatively it can be the cost of necessary investigations.
- Specific risks, which reflect the uncertainty in the assumptions used to estimate the cash flow.

A common ground between bankers and project managers is that specific risks should be borne by the party most able to control the event. As pointed out by Barnes,[20] in traditional contracts the contractor is assumed to have inspected the site and carried out his own site investigation. The employer does not then have a strong motive to do enough investigation to establish the effect of ground conditions upon construction cost. In the New Engineering Contract, the contractor is to assume the ground conditions will be as described to him in the tender documents. Thus if only a minimal investigation has been conducted, the prices may be based on a wrong view of the subsurface conditions and the employer's risk of later delays and cost escalation will increase. The employer should now be more motivated to do enough investigations, thus ensuring the contractor does not carry a large risk which he can do little to reduce. This reflects to a degree the American development of a geotechnical design summary report as discussed previously. It also questions the government policy of transferring ground risks to contractors through design and build road contracts.

The perception of what constitutes a substantial loss is not easy. Scriven[21] states that there is likely to be a difference between the project company and the banker. Bankers are likely to consider variations in excess of 10% on cost and time estimates as being substantial. Suppose that a risk analysis for a contract produces an estimate accuracy with a standard deviation of ± 7% of the total cost, the estimate has not included ground risks. If the risk associated with ground conditions is found to have a standard deviation of ± 12% of the total cost then the total risk is $(7^2 + 12^2)^{\frac{1}{2}}$ = 14%. Such a level of ground risk is of the right order of magnitude for many site investigations. In such circumstances, the ground can contribute 50% of the total risk to a project. It deserves more serious consideration.

---

[20] Barnes M. *The Role of Contracts in Management*. In *Construction Contract Policy - Improved Procedures and Practice*, Uff, J. & Capper, P. (Eds), Centre of Construction Law and Management., King's College, London, 1989.

[21] *Op Cit.*

Ground risks

In construction, the most serious effects from risk events are; failure to keep within the cost estimate, failure to achieve the completion date, and failure to meet quality and operational requirements. Risks and uncertainties need to be managed and accounted for, most importantly during the early stages of a project. The systematic management of risk consists of:[1]

- Hazard identification and assessment;
- Risk identification;
- Risk evaluation; and
- Risk control.

Ground conditions pose a real hazard to all construction. As construction projects are individual in nature the type of risk will vary. In evaluating ground risks there is usually insufficient objective data to calculate specific outcomes and resort has to be made to engineering judgement. Control of the risk is achieved by the provision of resources to investigate the hazard followed by design and implementation of the control measures.

Risk Analysis

The simplest form of risk analysis, Deterministic Risk Analysis, handles each risk independently of others and no attempt is made to assess the probability of occurrence. Probabilistic Risk Analysis is more sophisticated in that the probabilities and independence of identified risks are assessed. Risks are treated as variables having a specified probability distribution between upper and lower bounds, least and maximum possible values, around a most likely value: best estimate. Analysis is achieved by random sampling, such as Monte Carlo simulation, from each variable and calculations are performed

---

[1] Harris N., Herbert S. *Contaminated Land: Investigation, Assessment and Procurement*, ICE Design Practice Guide, Thomas Telford Ltd., London, 1994

many times. A number of uncertain activities can thus be combined to calculate an overall probability distribution for a project. This results in a most probable cost (or time) with estimates of the uncertainties associated with this value.

The effect of a change in a particular variable, or group of variables, can be investigated by means of a sensitivity analysis. In this values are calculated for each variable independently of others and the consequent effects are obtained. Thus the most critical, *i.e.* the most sensitive, parameter can be recognised; this is a valuable insight when managing a project and directing resources to best effect. In order to examine ground risks analysis has been carried out using the Computer Aided Simulation for Project Appraisal and Review (CASPAR) program. This models the interaction of time, resources, costs and revenues attached to each activity. It is based on a precedence network for a project and time related charges are handled by means of a hammock. The resulting cost model allows risks and uncertainties to be modelled with results presented in terms of parameters such as cost, Net Present Value and Internal Rate of Return.

The CASPAR program has been used to develop a risk analysis model to perform the following functions:

- To illustrate the sensitivity of various risk variables (including ground risks) on a project.
- To calculate the effects of uncertainty in ground conditions on the estimated cost and duration of a project. This provides an insight into the degree of adequacy of ground information on project feasibility assessments, planning and forecasting.
- To quantify the effect of work arising from unforeseen ground conditions on the cost and completion date for a project. This provides evidence of the consequences of inadequate investigations on the construction process

when 'unforeseen' events occur. These can lead to claims or compensation events.

As construction works, and associated ground conditions, are so variable it is not possible to quantify effects on a general basis, use has been made of case-study projects. With a project there is only one real event, what has occurred, the cost is known and operational times are complete. The model analyses various 'what if' scenarios to examine how the project could have been influenced by events other than the real case. The results can be compared with the reality to assess the significance of the calculations and to be better informed for future decision making.

Case Study : Tunnels

A tunnel contract involved five separate drives at the site, Figure 1. Ground conditions were reported from borings and consisted of mixed glacial deposits (sand, gravel, cobbles and boulders and clay) overlying sandstone bedrock. Sandstone was anticipated along parts of drive 1 and 5.

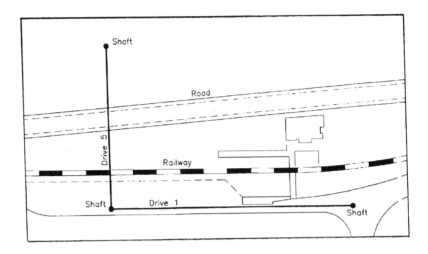

Figure 1: Site Plan A.

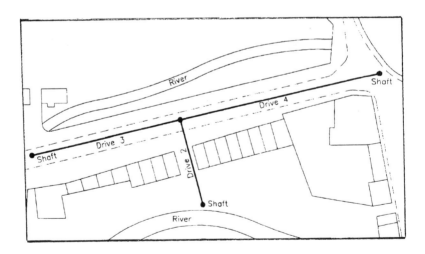

Figure 1: Site Plan B.

Construction : The contract sum was £440,000 but escalated by £159,000. All, but £10,000, of the escalation was attributed to ground conditions.

Drive 1:   accomplished with no serious incidents or failure

Drive 2:   successful

Drive 3:   major problems were encountered

- an unforeseen petrol tank was discovered at a previously dismantled petrol station not discovered by the site investigation. This was removed by a heading driven from a construction shaft.
- a water main was damaged. Delay was experienced due to excavation to repair the main.
- the tunnel encountered clay softer than expected and advanced at a slower rate than planned

As a consequence of the difficulties, the method of tunnelling changed to hand driving but at a slower rate and increased cost.

Drive 4:   hard sandstone, which was unforeseen, was encountered and productivity was poor and less than that planned. Costs increased as a consequence

Drive 5:   Completed without major incident, though progress was slower than expected

The project was modelled using 19 activities in the precedence network with costs and resources being supplied by the client. Risk variables were assigned to the ground conditions as follows:

- Unforeseen natural ground;
- Unforeseen artificial conditions, for example; underground services;
- Material failure;

- Machine failure.

The level of risk was assigned from a database of construction events on over 25 contracts involving nearly 100 tunnel drives. Subjective judgement was used to define uncertainty levels around each level of risk. Costs were assumed to be time dependent and thus when a hazard event occurs then cost variations are proportional to time variations. A sensitivity analysis, Figure 2, shows unforeseen ground conditions to be a very critical (i.e. the most sensitive) cost variable. The database on construction events shows machine failure to be more frequent than unforeseen ground but the occurrence of the latter is more costly.

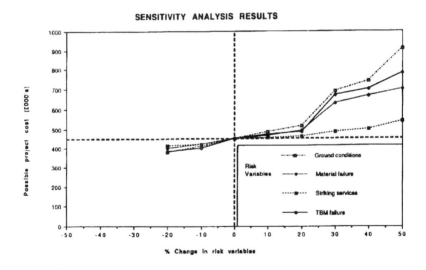

Figure 2: Sensitivity analysis

a) Ground uncertainty on tunnel cost and time estimates

Site investigation cannot eliminate all the uncertainties in ground conditions but the degree of uncertainty is influenced by the extent and quality of the site investigation, including desk studies and the like. For the analysis the following uncertainty levels were defined:

Uncertainty Level	Ground Uncertainty	Quality of site investigation Work
1	10%	Very Good
2	20%	Good
3	30%	Moderate
4	40%	Poor
5	50%	Very Poor

Table 1 : Ground Uncertainty Levels

The selection of these uncertainty levels was subjective. Later analysis of American data showed these levels to be reasonable. A proportional relationship was assumed between the ground uncertainty and variance from best estimate levels of project duration and cost. Thus a good quality site investigation provides reliable information such that major claims and delay from unforeseen ground conditions are avoided. Some variance, however, is possible but has a small effect on cost and time, less than 10%.

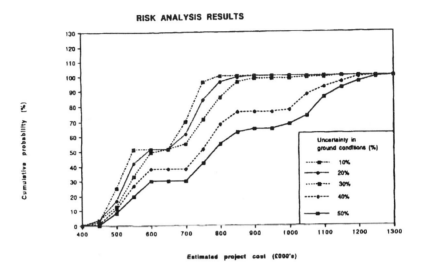

Figure 3: Effect of uncertainty in ground conditions on estimated project cost

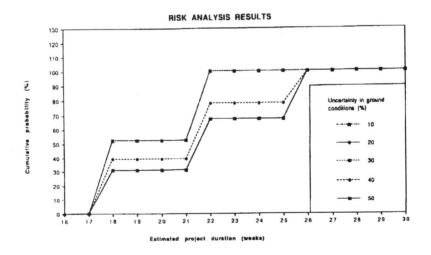

Figure 4: Effect of uncertainty in ground conditions on estimated project time

Figures 3 and 4 show the cumulative probability results for the contract cost and duration. The most probable value is associated with the mean - 50% cumulative probability. The limits of 0% and 100% represent extreme variation conditions and a more realistic appraisal is obtained from the 15%-85% probability range, that is extreme conditions are discounted. Table 2 summarises the data:

Level of Uncer -tainty	Mean Estimated Variation in Cost (£1000)	Range of Estimated Variation in Cost (£1000)	Mean Estimated Project Delay (weeks)	Range of Estimated Project Delay (Weeks)
1	100	40 - 270	0	0 - 3.5
2	130	50 - 310	0	0 - 3.5
3	160	60 - 360	0	0 - 3.5
4	300	70 - 590	3	0 - 7
5	330	90 - 500	3.5	0 - 7.5

Table 2 : Effect of Ground Uncertainty on Project Duration and Cost

The analysis shows cost and delay to be more likely as levels of site investigation deteriorate. In particular the trend is magnified for 'poor' levels, i.e. levels 4,5 and reliable estimates of project cost and time would not be possible. The quality of the site investigation is a real factor that can directly influence the reliability of cost and time forecasts when planning a project. Any investigation of less than 'good' quality is not reliable and should not be accepted.

b) Unforeseen ground conditions during construction

Assessment of the information for the contract led to the judgement that a relatively 'good' site investigation had been undertaken, as measured against present industry standards. The project was then modelled for four scenarios or possible site events:

- No unforeseen ground conditions, i.e. no extra work on any drive due to the ground;
- Unforeseen conditions on drives 3 and 4 caused 2 weeks and 1 week extra work;
- Unforeseen ground caused 1 week extra work in each of the 5 drives;
- Unforeseen ground caused two weeks of delay for drives 1, 3, 4 , 5 and 1 week for drive 2.

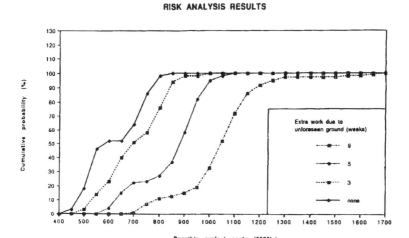

Figure 5: Effect of unforeseen ground conditions on project cost

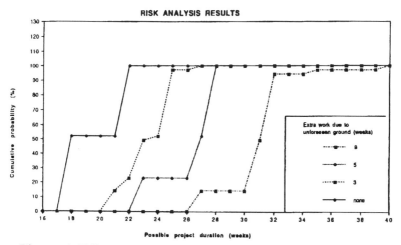

Figure 6: Effect of unforeseen ground conditions on project duration

The project cost variations and delay are shown in Figures 5 and 6 and summarised in Table 3

Unforeseen Delay Due to Ground (weeks)	Mean Likely Variation in Cost (£1000)	Range of Likely Variation in Cost (£1000)	Mean Likely Project Delay (weeks)	Range of Likely Project Delay (weeks)
none	130	40 - 270	0	0 - 3.5
3	250	110 - 370	5	3 - 6.5
5	450	200 - 500	9	4.5 - 9.5
9	600	410 - 700	13	9 - 14

Table 3 : Variation in Project Cost and Duration due to Unforeseen Ground Delay

It is to be noted that even with no unforeseen delay due to the ground there is still a risk of cost escalation and delay from inherent ground uncertainty and other risk variables, such as striking underground services, material or machine failure. Unforeseen ground can be seen to have a significant effect on project cost and duration. For example, scenario 2 (3 weeks delay) suggests the mean likely cost increase to be 57% of the tender value (£440,000) with an uncertainty range from £110,000 to £370,000 or 25% to 84% of planned costs.

c) Model and Reality

The project was completed with no significant delay but costs increased by £159,000, due mainly to ground conditions. Most cost overrun occurred in drives 3 and 4 and delays were avoided on site by increasing productivity and resources, and hence increased cost.

The risk analysis produced a close agreement between the likely estimated variation in cost (£130,000) and the actual increase (£149,000) due to ground conditions.

The model analysis shows that for tunnels a high quality of site investigation is required if cost uncertainties from the ground are to be reduced to acceptable levels. This supports the analysis of tunnel data from America. Even if ground conditions are proved to adequate levels, tunnel contracts still have inherent uncertainty from other factors such as machine failure. It is probable, however, that tunnel methods would be well designed to suit the ground if there was a high quality site investigation and thus the mechanics of tunnelling would be more reliable and less prone to failure. Ground risk modelling is a valuable tool in planning a project and gives confidence to the reliability of cost and time estimates for the works. In the case study contract, the tunnels used technology previously untried by the client and higher levels of contingency were allowed. Even though the cost over-ran, the project was considered to be successful and risk modelling proved valuable to the exercise.

Case Study : Industrial Development

The study concerns an industrial development on a 'brownfield' site. The land was previously occupied by steel works and had been reclaimed for light industrial use. The site was sold to a development company for the construction of a large industrial warehouse and offices designed to the heaviest industrial loading standards. Before construction, agreement had been reached with the industrial client on rental terms and building specification. A fixed commissioning date was contracted for to meet commercial targets and any delay was subject to penalty by the developer.

a) Ground Conditions

It was known that the site contained underground obstructions from old foundations amongst other things. These were detected by desk study and exploration. The site purchase price contained an allowance for these obstructions and no additional pre-contract surveys were commissioned: the main investigation carried out related to light industrial development. The exploration consisted of grid borings at 50m centres across the site. Contours of fill thickness were determined with reasonable reliability, the fill was underlain by a thick layer of mixed Glacial Deposits (sands and clays).

b) Construction

Foundations were designed as driven steel piles with locations chosen to avoid obstructions. As piling started, many piles could not be driven due to obstructions that were not foreseen, The site was excavated to clear the problem and old foundations, walls, etc were encountered at locations not detected by the investigations. These major earthworks cleared the site but caused a delay in the planned site clearance period of 2 weeks to 10 weeks total. More resources had to be brought to site to reduce the delay produced by this critical activity and to meet the scheduled commissioning date. The total planned cost for the contract was £4.3 million, the additional 'unforeseen' ground costs were estimated to be £900,000, ie. about 20% of the contract value.

It has been reported, but not confirmed by the author, that an industrial archaeological study of past use identified industrial buildings which pre-dated the steel works. The obstruction from the steel works were known from the decommissioning records, but the obstruction from previous constructions were not known. The importance of desk studies into site history and geological conditions cannot be over-emphasised: they are essential, and are obtained at relatively modest cost. The risk analysis was similar to the tunnel case study reported above and details are given by Alhalaby and Whyte.[2] A sensitivity analysis showed ground uncertainty to be a

---

[2] Alhalaby N.M.H., Whyte I.L. *The Impact of Ground Risks in Construction on Project Finance. In Risk and Reliability in Engineering*, B.Skipp (Ed.)

significant cost variable when compared to other events such as delays in planning permission, land purchase and design. Modelling the effect of site investigation quality on planning the contract showed the following results for estimated cost escalation on the contract sum:

Quality of site investig- ation	Mean Estimated Cost Variation (£1000)	Range of Estimated cost Variation (£1000)
Very Good	65	10 - 135
Good	140	25 - 345
Moderate	235	40 - 485
Poor	380	60 - 675
Very Poor	465	95 - 855

Table 4 : Effect of Ground Uncertainty on Project Cost Escalation

Proceedings of an Institution of Civil Engineers Conference, Thomas Telford Ltd., London 1994.

The effect of delay during site clearance operations, as occurred on the site, produced the following results :

Unforeseen Ground Delay (weeks)	Mean Estimated Cost Variation (£1000)	Range of Estimated Cost Variation (£1000)
none	140	25 - 345
2	360	230 - 600
4	660	400 - 820
6	820	590 - 1030
8	1070	780 - 1370

Table 5 : Effect of Unforeseen Ground Delay on Project Cost

Note: In developing the unforeseen ground delay costs it was judged that the site investigation commissioned was of a 'good quality'.

It is of interest to note that the actual 8 week delay produced an estimated cost addition of £900,000, well within the probable range of variation and not too dissimilar to the mean estimated cost of £1,070,000 from the model. This provides some confidence to the validity of the modelling process.

Discussion on the Ground Uncertainty

The modelling process has been applied to two other contracts, an industrial works and a new industrial plant. The results produced are similar to those reported above. The following can be noted:

- Many projects are designed and planned before a site investigation and exploration are commissioned. There are time and resource limitations to investigations and problems of communication. Checking the adequacy and accuracy of ground information is the exception rather than the rule. Site investigation practices vary from good to very poor, good practices are observed generally in organisations with a recurrent development budget. There is a lack of knowledge on the financial risks and uncertainties as a consequence of less than adequate investigations.

- Data for case study analyses are difficult to collect, due to their often sensitive and commercial nature and problems of confi-dentiality.

- At the construction planning stage, uncertainties in ground information lead to unreliable cost and time estimates. The degree of unreliability increases dramatically for investigations that are less than 'good' in quality. The cost of an adequate investigation is relatively modest, for example, a low quality site investigation for the industrial development case study could cost as little as £10,000, and a very good site investigation could cost around £40,000 - £60,000: around 1% to 1½% of the estimated cost. The benefit of a good investigation can be measured not only in the reduction in mean probable cost escalation but also in the associated range of financial uncertainty. In the case studies, the reduction in cost overruns obtained from better information on the ground is measured in terms of £100,000's. The return from quality work far exceeds the cost increase in improving the quality from 'poor' to 'very good'.

- The effect of unforeseen ground conditions being encountered during construction, consequent upon less than adequate investigation, has been modelled in terms of cost increase and delay. In the case studies, unforeseen conditions resulted in cost increases measured in £100,000's with high levels of uncertainty on what the cost variance could be. Should an adequate investigation be capable of detecting problematical ground conditions, that is the events are foreseeable, then the modest cost of the investigation again produces an order of magnitude return in the level of financial risk to the client and bankers.

- Analysis of costs produced by delay showed that a 1 week delay can incur an escalation in excess of £100,000, a figure far in excess of an site investigation cost. It would seem reasonable, therefore, for a client or banker to estimate the cost of a 1 week delay to his project and allow this as an upper estimate of the cost of investigation necessary to avoid such a delay. This is a more rational basis than a notional percentage of estimated project cost.

- Ground conditions are inherently uncertain and there will always be an element of risk no matter how much is spent on investigation. The important feature of an adequate investigation is that the risk is probably reduced to acceptable levels. It is evident, however, that such levels are not prevalent in the industry and that change is required. For change to occur, investment is needed in resources of time and money. The industry has recognised the need for change but, as indicated by Jardine and

Johnson,[3] there is little success to report in achieving the change. As resources are required and these are controlled by promoters and their bankers and financiers perhaps they will drive the engine for change. After all, it is their money that is at risk.

- Procedures for good practice have been reported by CIRIA[4] and the Institution of Civil Engineers.[5] A strategy to account for risks has been presented by Tonks and Whyte[6]

- An early study on ground risks[7] reported a case study on a supermarket development. This showed that a proper investigation can lead to increases in cost estimates from initial values, The new estimate is, however, more reliable and allows contracts to be efficiently managed and financed. This leads to better cash flow calculations. Promoters do generally want to know in advance what their costs are likely to be and dislike estimates which suffer subsequent escalation and claim inflation. Also, overly conservative estimates may mean that the client borrows or locks up more money than is necessary.

---

[3] *Op cit.*

[4] Uff J.F. & Clayton C.R.I., *Op cit.*

[5] Inst. Civ. Eng. *Site Investigation in Construction: 4 reports, Op cit.*

[6] Tonks D.M. & Whyte I.L. *Project Risks and site investigation Strategy.* In *Risk and Reliability in Engineering*, B.Skipp (Ed), Proceedings of an Institution of Civil Engineers Conference, Thomas Telford Ltd., London, 1994.

[7] Whyte I.L. & Peacock W.S. Site Investigation and Risk Analysis, Proceedings of the ICE, Civil Engineering, May 1992, 74-82

Conclusions

Site investigation practice in the UK has developed from a competitive contractual background. The industry has suffered from work of indifferent quality, low price levels and insufficient resources. The problems are recognised within the profession but there seems to be little evidence of change in practice. Evidence on contracts in the UK, and America, during the 1980's began to quantify the consequences of inadequate investigations in terms of cost and delay. Procedural changes to avoid underground problems were developed in America and are worthy of consideration in Britain. Likewise, procedural changes are occurring in the UK with some clients passing the ground risk contractually to the contractor. Bankers and financiers review projects in terms of security and finance. Major concerns are cost overruns and time delays. The perception of what constitutes excessive risk is ill-defined but could be of the order of 10% on cost and time estimate. Such variance can readily be accounted for by unexpected ground hazards.

Ground risks have been researched using a probabilistic risk analysis of selected case study projects. Such analysis confirms ground uncertainty to be a sensitive factor in cost control. It also confirms that additional resource to produce an adequate survey results in significant reductions in both the probability of cost escalation and the range of financial uncertainty associated with the ground. It is suggested that the cost of one weeks delay to a project be calculated, and this figure used as an estimate for the ground investigation costs necessary to avoid such a delay. Bankers and financiers provide the capital for development and receive the consequences of ground hazards appearing unexpectedly on a site. It is in their interest to demand not only value for money but also quality. If the resources are available, then it is up to engineers to put into practice the principles of change to good standards that they have long advocated.

12. Financial Risks in Construction

David Richmond-Coggan

Synopsis

This paper seeks to define financial risk in the context of construction. The sources of and means of accommodating financial risk are considered, primarily from the client's perspective.

Introduction

In the modern management of construction, project finances are no longer the sole preserve of the quantity surveyor. It is becoming recognised that there is a need for the financial management of a project to be the responsibility of someone with a wider view of the project as a whole: the project manager. The project manager should be close enough to the client to have a sufficient understanding of the overall objectives of the project, such that he is able to make judgments and offer sound advice to the client on value within the project. Although the concept of value extends beyond just value for money, it is this element of value that most affects the project finances.

What is Financial Risk?

Risk is generally considered to be the chance of a bad consequence or loss, in effect, the exposure to mischance; and financial risk is generally considered to be that associated with revenue or expenditure. Thus, financial risk will be the chance of a bad consequence relating to revenue or expenditure. The balance between revenue and expenditure, either at a point in time, such as that reflected in cash flow projections, or at the end of an accounting period, such as at the end of the project, is the measure of the financial success or failure of the project. Money is the most tangible measure used when determining the outcome of a project. If quality has been poor, or if works are completed early or late, the detriment to the project is usually measured in terms of money. Thus it may be said that all risks are financial risks in as much as they all have some financial effect.

What is construction in the context of Financial Risk?

Construction is generally understood to be the process of building[1] something, whether it be a dam, a motorway, an off-shore oil platform, an office block, shopping centre or house. Where does this process start and where does it finish? Does the process start with a gleam in the eye of a potential client, with the appointment of an architect or engineer to prepare an option study, with the decision to commit funds to the design of a particular option or the appointment of a Contractor to start building? Does the process finish with the issue of the Certificate of Practical Completion, the issue of the Final Certificate at the

---

[1] For convenience and brevity throughout this paper, the term **building** is used to describe anything being constructed. It is not intended to limit the scope to the items of construction normally described or known as buildings.

end of the Defects Liability Period, or at the end of the Design Liability Period? Does the post-project appraisal form part of the construction process and, with the increasing desire for energy and maintenance efficient buildings, does the construction process in some sense continue into the occupied life of the building? The answers to these questions will have a significant bearing on the financial risks to be discussed, but the answers will be very different for the various parties involved.

Viewpoints of Financial Risk

There are many parties involved in the modern construction process. These are:

- The Client;
- The Financier or Provider of Funds;
- The Project Manager;
- The Designers;
- The Contractors;
- The End Users; and
- The General Public.

The boundaries between these parties may be quite blurred on many projects, but it is still useful to consider the fundamental functions that are carried out and that carry exposure to risk. As soon as a potential client embarks on expenditure for professional advice, or in respect of his own time, on an idea for constructing something, he is exposing himself to financial risk. His exposure to risk will continue at varying levels whilst he decides what to build, whilst the design process is underway, during the construction phase, and ultimately when he puts the building into use, either for himself or with others

The provider of the funds for the project exposes himself to financial risk when he commits funds to the project. His exposure to risk will continue until those funds are fully repaid or written off. The project manager and the designers expose themselves to financial risk when they expend money on tender pre-qualification work, or when preparing an appointment bid. For those appointed, this exposure continues until all liability under the contract is ended.

Contractors are exposed to financial risk when they expend money on pre-qualifying works or in tendering for a contract. Similarly, their exposure continues until contractual liability ends. For contractors the risks to which they are exposed will be greatly affected by their experience of the work involved, the market conditions at the time of their appointment and during their contracts, and the form of contract under which they are appointed. The end users may unwittingly be exposing themselves to future financial risk as soon as the form of the building is decided: decisions will have been taken that affect the cost of their use of the building. Their exposure to risk may be considered as continuing while they continue to occupy the building.

The involvement of the general public in a construction project can take many different forms. They may be the ultimate financiers, via taxation, of a public project. They may be the end users for example of a motorway or a public library. They may be adversely affected by the building process itself, either from physical risk of harm or from inconvenience. Indeed, they may be affected by the environmental impact of the building.

Sources of Financial Risk

Following the traditional procurement path for a construction project, the project goes through the following stages:

- Option and feasibility studies;
- Development of the Project Brief;
- Development of the Design;
- Tender for the Construction;
- Construction;
- Commissioning; and
- Taking over and occupation

Under different procurement paths, these stages can overlap, become confused or occur in a different order. This does affect the exposure to financial risks, the type and size of risks, the parties that they affect, and the methods to deal with them. Sources of financial risk may be completely external to the project. They may arise from Government policy, changes in priorities during critical stages of the project, changes in the tax regime, changes in the anticipated rate of inflation or by interest rate fluctuations. For projects near the limits of viability, these changes can be severe. All those responsible for the project must be aware of their effects and the tolerances acceptable for the continued well being of the project.

The Stock Market Value for companies in the private sector can have a significant impact on their ability to finance projects. This facet may not be under their control if the Stock Market is generally weak or falling. The economic cycle of boom and bust has a major impact on the construction industry which finds itself used as an informal economic regulator. Shortages of labour and materials affect the ability to complete work to programme, while a glut of work causes costs to rise uncontrollably.

Many of the sources of direct financial risk for the client are found in the early stages of the project. The risks associated with the overall suitability and efficiency of the completed building for its intended purpose stem, to a very large extent, from the process of developing the project brief and the choice of professional advisors. If the client is experienced in the procurement of similar projects, this process may be relatively straightforward and carry low risk. For example, a high street supermarket chain will have a highly developed brief already prepared, based on previous experience, and will probably commission professionals who have worked on a similar project and who will therefore have a good understanding of the aims of the client and the project.

However, if the supermarket client decides that the market has taken a significant shift, or if the project brief needs significant change, the risk that the project brief will not clearly impart the fact that the client's expectations. Similarly, if the client decides that he can reduce the cost of professional advice by inviting open bids for the provision of the service, it is possible that the lowest bid may be submitted by a firm who will not provide the expected service. The firms with experience of similar projects will bid on the basis of past experience and may be more expensive. The client must decide whether the risk of paying a higher fee to a firm with whom he has a comfortable relationship, and whose output is known to be satisfactory, is likely to be higher than the risk of employing a firm who may need more time-consuming input from the client and whose output may not be entirely satisfactory. There will be occasions when each of the different decisions will be preferred. Even if the client is experienced in the procurement of similar projects, risk that external constraints may affect the development of the project brief is ever present. For example, decisions may be more affected by policy on expenditure than on the most efficient whole life cost of the building.

If the project brief is not successfully transmitted to the professionals, various financial risks may occur. The project may develop in the wrong direction and require additional input of time and effort from the client to bring it back on course. This may generate claims for additional fees from the professionals and may also cause a delay in the programme. If the client fails to realise that the project is developing in a direction he does not want it to and corrective action is not taken, a contract may be let and construction work be at an advanced state, or even completed, before the problem is recognised. The financial effect of undertaking remedial work or re-work together with the delays involved, or of using a building that is not designed efficiently for its intended purpose, may be very significant.

The risks to the financier arise in similar ways, but usually with less direct involvement in the project. The financier's exposure is, however, somewhat different. The financier is potentially exposed to risk both from the client's financial stability, and from the intrinsic value of the building, possibly to another user. The role of financier is changing with the development of Build, Operate and Transfer (BOT) and Build, Own, Operate and Transfer (BOOT) projects and of the Government's Private Finance Initiative (PFI). The financial risks to which the financier is exposed are changing. In these circumstances, the financier is much more closely involved in the project as a member of the project team.

Much of the financial risk to the consultants and contractors arise from the method by which they are appointed. Generally they will be appointed on the basis of some form of competition. The essence of winning in competitive bid situations is in balancing the estimated cost of carrying out the work specified in the bidding document with the estimated going rate for the work in the market place. Under present market conditions, the going rate will almost always be less than the estimated cost of fully satisfying the bid specification. Most experienced clients are aware of this situation and expect to have to manage the financial risks that this situation exposes them to. The consultant or contractor will inevitably be looking for opportunities to claim additional income: less experienced clients will not be aware of this situation. The financial risks to the end users are less obvious. They will depend to some extent on their relationship with the client. The risks are likely to relate to the cost in use of the building and the cost of occupation.

For a conventional office block or factory, the rent, notional or otherwise, will relate to the need of the client to recover the capital expenditure at the required Rate of Return. The maintenance, later refurbishment and running costs will depend upon the quality of the materials used, as well as the quality of the design; any improvement sought has an inevitable capital cost. In the private sector, a company building for its own occupation or use will display a close relationship between the client and the end user. The end user is likely to be involved in the early development of the project brief and thus able to ensure that it satisfies his aims and objectives. There however may be conflict between the demands on the capital and revenue budgets: when the capital cost of building for low maintenance or low running costs is being calculated there is likely to be an inherent flexibility to make decisions to provide the lowest overall cost. In the public sector there has frequently been a separation between the client and the end users with capital expenditure and occupation costs coming from different budgets. Treasury guidelines state that, *inter alia*, minimum lifetime cost and low maintenance should be designed for, it is extremely difficult in practice to justify increasing capital costs where the budget is already under pressure. In these circumstances, there is significant exposure of the end user to financial risk.

The general public are unlikely to be directly exposed to financial risk until the construction starts. Fundamental decisions about the project and its design may, however, affect their later exposure. The greatest exposure during construction is from health and safety aspects which, if an accident should happen, would have a financial effect. Inconvenience during construction such as closed roads, travel delays, restricted access to shops, causes a waste of time which has a financial value. The impact of the building on the environment represents another financial risk for the general public. The examples of outfalls from sewage treatment plants, CO_2 emissions or radioactive waste from nuclear facilities are well known, as are the effects of CFCs on the atmosphere which underline the need to use materials that do not use CFCs in their manufacture.

Dealing with Financial Risks

Financial risk is an element of construction which pervades all elements and stages of the process. Since a risk is essentially something unknown it can never be removed completely, but if sound management systems are put in place risk can successfully be dealt with. Dealing with financial risk is only one of the management systems available to the construction industry. It is recognised that a client and his professional advisors have the greatest scope for overcoming problems at the earliest stages of a project. As the project progresses, so the scope for overcoming problems without serious effect on the programme, budget and quality of the project are reduced. Management to deal with financial risks should be put in place at the earliest stage of a project, but must be closely integrated with the overall management of the project. A single point of responsibility for the management of the project ensures that all aspects of the project are considered. Risk, value, quality, time or cost cannot be considered in isolation from one another, they are all closely inter-related.

The appointment of a competent project manager at the earliest point in time with responsibility to manage the project through to completion, is the first and most important step in dealing with financial risk. If the client has the necessary resources and skills in house, the project manager can be an internal appointment. If the client has no such resource available, or considers it advantageous to have more independent advice, then an external professional should be appointed. Whoever is ultimately appointed as project manager should be appointed on terms which allow him the opportunity to fully understand the aims and objectives of the client. To carry out this function successfully, the project manager will need access to confidential information and will therefore have to have the complete trust of the client. In addition the project manager should have the necessary skill, expertise and experience to help the client translate the aims and objectives for the project into a brief which successfully imparts the requirements to those employed to implement the works.

The client, with the assistance of the project manager, will have to make early decisions about the process of procuring the project. The design may be carried out ahead of construction or in parallel with it. The designers may be employed directly by the client or be part of a complete construction package. The financial risks likely to affect a project will be carried by different parties under these options, but the client should be aware that transferring risk to another party is likely to have a direct cost. To assist with the decision making at this and later stages of the project, there are techniques of risk assessment which should be used. It is not necessary, particularly on small projects or at the early stages of projects when good information is not yet available, to employ high powered computer software for the analysis of the financial effect of risks. The most useful part of a risk assessment exercise is the identification of the possible risks by the project team and consideration of the possible management action that could be implemented to minimise the risk.

Identification and evaluation of the risks represents the first stage of dealing with them. The next stage is deciding on suitable management action to counter the risks identified and to set in place a systematic procedure for dealing with the risks that become apparent during the project, but which have not been adequately identified earlier. The management action may take a variety of forms: transfer or sharing of the risk by means of the contracts entered into, including insurance arrangements; further investigation of the risk for example by ground investigation or market research; retention of the risk where this is reasonable; for example, the risk of weather delaying the works. Decisions on sharing, transferring or retaining the risks should be supported by analysis of the effect on the overall finances of the project of the alternatives, by considering the sensitivity of the finances to the options available.

It is a fundamental principle that risks are best carried by those that are best able to deal with them. It is important to remember that risks may be transferred by various mechanisms. The most cost effective time to transfer a risk is at the stage when sufficient is known about it to reasonably estimate its magnitude, but early enough to allow time to take necessary action to mitigate its effect. There will be some risks that are most cost effectively carried by the client. However, by accepting these risks, the client will deny himself cost certainty at an early stage of the project.

In conclusion, the process of identifying, evaluating and dealing with risks is a continuous process throughout the project. As the project progresses, some risks are overcome or do not occur, and new ones become apparent. Better information becomes available to assist in the identification and evaluation of the potential risks that may affect the project. A formal reconsideration of the risks affecting the project should be carried out prior to each major decision point, just as a formal reconsideration of the value of the project should be carried out at these key points.

13. Control of Time in Construction

Gary France

Synopsis

This paper offers practical guidance in the control of time in construction in terms of the author's Ten point plan.

Timing of the "whole" process

Construction is not merely the stage during which a new building is built. "Building" comes at the end of a whole series of stages, events, decisions and activities all of which make up the total process of creating a building. Most of the control of the construction process is gained by the things that are done way before construction starts. More time is lost through the wrong decisions being made (or even worse, not even considered) before construction starts than after. Therefore it is essential to consider all parts of the process to achieve the appropriate time allocations between the feasibility, design, procurement, construction and occupation stages. Sufficient time must be allowed for the client/end-user/occupier to play their part in the process. Development of the brief and signing-off of the design are both critical: adequate periods must be allocated for these activities. Perhaps the biggest single cause of delays to construction projects is in finalising the various legal agreements that need to be put in place. What should be a relatively simple process often takes an inordinate amount of time and control of the timing of legal agreements should always be made a priority.

Get the "constructor" involved as early as possible

Traditionally, the time spent on construction projects was in two distinct halves: design and then implementation. This manifested itself by creating two separate teams, one set of people to design and a totally different set of people to construct. The result was little integration, little co-operation and little benefit. This separation of roles happens in virtually no other industry. What is needed is for the designer and constructor to work together and to help each other with advice where and when it is required. Appointing the "constructor" as early as possible encourages co-operation. It is of tremendous benefit to the control of time in construction if the construction strategy is developed in parallel with the design strategy, for then the designers and the constructors are working together, respecting the needs of each others complementary skills to arrive at a solution which can be built economically and quickly.

Choose your procurement route carefully

The effectiveness of controlling time in construction is largely affected by the procurement route chosen. Some procurement routes offer little opportunity to control time in construction. These are often the JCT fixed price type contracts that stipulate a start and a finish date and little else. All aspects of control of time are handed over to the main contractor and the rest of the project team have little or no influence on how and when activities are undertaken. On any project, time is often the most valuable asset and it must be properly managed. The procurement route selected must provide the client with the ability to make changes without undue time penalty. For the client who knows exactly what he wants from the outset, and is confident that they will make no changes, then fixed price contracting is recommended. For most projects this is not the case. What is therefore needed is a contractual arrangement that allows the various parties to work together with the sole purpose of working in the clients' best interest. I would advocate Construction Management as the most appropriate route for the majority of projects.

Building effective teams is critical

The construction industry is notorious for its tug-of-war relationships, but it need not be so. Control of time in construction can be greatly enhanced when all of the parties involved are working with a common goal, ensuring that they pull together and in the same direction. This concept of common goals is fine, but how is it achieved? First, it is necessary to create and then manage an environment that truly embraces teamwork rather than to stand back and point the finger of blame when things go wrong. Creating a team in which an open, trusting atmosphere prevails should be the top priority at every project's inception. Hopes, wishes, aspirations, concerns and needs are all out on the table. This creates a common ownership where everybody understands the project goals and, therefore, begins to understand the overall needs of the project, not just their part within it. They become stakeholders in the project and are therefore committed to working towards and then sharing in its success.

Control of risk through analysis and management

Risk analysis and risk management are sometimes regarded as modern day management functions but, in fact, they have been used for hundreds of years. We all assess risk all of the time and use risk analysis in our everyday work on construction projects. However, risk analysis has in recent years been over complicated by boffin-based computer systems that, by using such methods as Monte Carlo, can tell you that given a 20,000 simulation test of your project, that you have got a 17% chance of completing either 3 weeks ahead or behind programme. Great. Very re-assuring. Totally meaningless!

Mathematical risk analysis is often referred to as the quantitative stage and is the one that most people rush to. I advocate concentrating on the qualitative stage and this is the part that is by far the most useful. To assist in the control of time, all projects would benefit from undertaking a very simple form of risk analysis. This is in four steps:

- think of no more than 20 risks, that if they occur, will delay the project;
- give each a mark out of 10 for the likelihood of it occurring (where 1 is low, 10 is a high likelihood);
- then consider the impact to the project should such a risk actually happen. Again, award 1 mark for a low effect and up to 10 for a high effect;
- multiply the mark for likelihood by the mark for impact to arrive at an overall risk factor.

Clearly, those risks with the highest risk factors present the greatest threat to the project. This is risk analysis at its simplest, and most effective. Repeat the exercise each month to monitor and to track how the risks to the project are changing. The management of risk is now far more targeted, as the team can recognise, and then do something about, reducing the likelihood or impact of the major risks.

Pre-engineer the project

On any project, the pre-construction period is critical. Work done at this stage sows the seeds for the construction activity that follows; get the pre-construction work wrong and your harvest will be poor, get it right, and you will reap the benefits later. In essence, pre-engineering means setting up the managerial aspects of the project before going anywhere near the site. This ensures that the construction phase is solely about implementing already agreed plans, not still making them. Pre-engineering means that nothing is left to chance. Activities undertaken as part of pre-engineering include:

- setting up the project;
- working out sequences, methods and timings;
- agreeing with the designers the most appropriate methods and designs;
- being innovative in approach, but not by injecting unnecessary risk;
- splitting the work into packages;
- interface management;
- working out how the project will be bought;
- looking at logistics, access and materials handling;
- value engineering and value management;
- risk analysis and risk management;
- life cycle costings;
- surveys;
- talking to the buildings neighbours, local authority, police etc;
- developing a safety plan and a fire plan;
- building trusting, working relationships with all parties involved in the project.

Pre-engineering work is essential and must not be rushed or compromised. Without it the project will invariably suffer, leading to delays. Pre-engineering clearly needs the earliest possible involvement of the constructor or construction manager.

Involve trade contractors early

As with the constructor, construction manager or main contractor, the early involvement of the specialist trades is highly desirable. The construction industry is so huge that it is ridiculous to think that any one person knows everything. Therefore, by soliciting early advice, the control of time in construction can be enhanced, because you will have the knowledge that the design being undertaken, or the plans being made, are realistic and achievable. The advice of specialist trades can also be an advantage by those trades promoting new, faster or more economical ideas. Specialist trades contractors are invariably only too pleased to be involved early in a project even on an unpaid, no promises basis. Given the opportunity to be included on a tender list is reward enough for most specialists. There is an added benefit to this early involvement. Any specialists whose advice or help was taken in arriving at a design solution for a project will, provided they are later appointed, take shared ownership of that solution and will therefore want to do everything possible to ensure its success. The key to working with specialist trades, no matter what the stage of the project, is to recognise their expertise by giving them direction, not directions.

Fast build, not fast track

Perhaps the greatest affect that the control of time in construction can have is whether the project should be "fast-tracked". Fast-track is generally taken to mean over-lapping the design with the construction to achieve the earliest possible start on site date. Whilst the advantage is of starting earlier on site, the potential disadvantages are many and onerous. Fast-track contracts put the designers and the client under tremendous pressure. All of the thought processes are cut short. A great deal of potential flexibility is lost in the remaining design period, because the construction has already started. For example it becomes very difficult to make changes to the design. I do not believe in "fast-tracking". I prefer fast-build, where more time is taken initially: one does one's thinking first, and not on the run.

The design is thought through properly, giving the designers the time that they really do need to design out problems and time is used to manage out potential areas of risk.

Interface Management

Recognising and managing the various interfaces on a project also helps to control the overall time aspects, mainly by ensuring that things don't go wrong and hence cause delays. Time can be saved by ensuring, for example, appropriate tolerances.

A +/- 2mm metal window tolerance is no good if the brickwork reveals into which it is supposed to fit is +/- 5mm on each reveal. Both the brickwork and the window may be within tolerance, but the window still might not fit and time will be lost while this is rectified. For example, in the design of the structural frame, the engineer needs to liaise with the architect for form and layout, the M&E engineer for the effect of services design upon storey height, the cladding designer for slab edge loadings and fixings, and the piling trade contractor for foundation design. These discussions happen at different times and almost certainly on many different occasions. These interfaces all need to be managed in such a way that the design process is smooth and unencumbered by any missed communications between the key parties.

Trade contractors need to be aware of the other trades they will have to work with. One useful medium for this is the interface register which lists all those trades an individual trade contractor will have to recognise, meet with, design for, co-ordinate with and work next to on site. When "buying" a project in packages it is vital that there are no gaps between any two packages. By not buying say, the flashing between the cladding and the roofing, this can lead to lengthy delays in achieving watertight. The sequencing and timing of the actual construction clearly has many interfaces that need to be managed. The ability to recognise these and programme the works to respect how each trade contractor meshes with other trades is important.

Set realistic, understandable and achievable key dates

Don't waste effort and expense by trying to achieve the impossible. Understand what is realistically achievable and ensure that everybody works towards it. Set understandable key dates and publicise them widely. Do not let it be known, for example, that by week commencing December 16 the first floor slab should be concreted; that the first-second storey walls should be reinforced; and that 25% of the columns should be cast. Just simply say that the target is to complete the first floor slab before Christmas. This is far more motivating. Keep key dates simple and easy to remember. Publicise key dates widely; on site notice boards, attach a sheet of key dates to the back of minutes of meetings, put a large notice in the site reception showing what the key dates are. Inform to motivate is the key!

14. Major Infrastructure Projects and their Environmental Impact I *

Helen Payne

Synopsis

This paper seeks to identify the particular risks which may impinge on the construction of a major infrastructure projects. The effect of environmental risk is examined in the context of the project's long-term benefit.

Introduction

Construction of major infrastructure projects can be seen as one of the means of meeting the needs associated with population and industrial growth, such as water sanitation, power, transport and other infrastructure, and with meeting environmental concerns such as waste management and pollution control facilities. Investment in such infrastructure is clearly one of the prerequisites to continued growth.

Improvements in infrastructure and enhanced quality and reliability of supply can be an important factor in attracting new foreign investment and encouraging the growth of local industries. Trade and industrial growth can be inhibited by inadequate road and rail networks, airports, ports and other

---

* Based on the author's paper which won the 1994 Hudson Prize

transportation links as well as interruptions to telecommunications caused by power shortages. Investment in transportation and energy infrastructure can help in reducing the costs of distribution and the interruption of businesses through nower failure.

There has been a growing trend in recent years both in the United Kingdom and overseas towards reduced public spending on infrastructure projects. This has been demonstrated by governments or their agencies placing major projects into the private sector, rather than staying in the traditional domain of the public sector by using different procurement and contracting strategies to those previously employed. For example, Part I of the New Roads and Street Works Act 1991, which came into effect on 1 November 1991, enables private promoters to enter into concession agreements with the Secretary of State to finance, build and operate new roads and to charge tolls. Although not covered by the 1991 Act, the new Design, Build, Finance, Operate (DBFO) schemes for roads envisage a similar concept.

UK government agencies have been involved in projects which have been built and operated by a private company, but financed largely from public sources, such as the Manchester Light Transit Railway project which required invited bidders to tender for a concession to build and operate and maintain the system with ninety-five percent of the costs provided by the promoter to be recouped during the operation and maintenance period from revenues generated. Whereas in the Channel Fixed Link project, finance was procured by a private consortium of lenders and investors without any financial backing from the governments of France and the UK.

Inherent risks

The development and execution of any major project can often be a difficult and uncertain process. Risk and uncertainty are inherent in all construction projects, and the assessment of risk and reward is clearly fundamental to any venture. Risks need to be identified, appraised and allocated through a risk management structure which addresses all those risks over the life cycle of a project. Promoters and developers can be exposed to risks throughout the life of a project: failure at any one of several stages of the project, failure in the later stages of the project when considerable amounts of money have been expended in development costs; and failure of the project to generate returns or the opportunity to recover costs.

By identifying risks at the appraisal stage of a project a realistic estimate of the duration and final costs and revenue of a project may be determined. With concession arrangements, a structure in which not only the issues as between the principal and the promoter are addressed, but also those issues associated with construction, operation and maintenance, finance and revenue packages should ideally be used.

Risks associated with major infrastructure projects can be considered on two levels:

- **project risks** which are to some extent controllable and relate to construction, operation, finance and revenue generation; and

- **global risks** which are generally not controllable by the parties and are outside the project, such as are associated with political, legal, commercial and environmentnl factors. Risk characteristics of infrastructure projects differ on a project by project basis, but broad categories of risks can be distinguished.

Construction Risk

Whether the project can be built on time, to specification and within budget are usually referred to as completion or construction risks. The degree of this risk can vary between different types of infrastructure project. For example, it could be of considerable importance in relation to the design of a challenging bridge if there are doubtful ground conditions, although perhaps not so high in construction of a conventional motorway. Similarly re-routing, or replacement of habitat where a project encroaches on a site of particular scientific importance may have to be taken into account. Linked to the completion risk, other major construction risks are the type of technology adopted and the terms of the construction contract. A new and hitherto untried transportation system; for example, an unproven city centre monorail system, can pose a high risk and is less likely to secure debt funding on a project finance non recourse basis.

Other risks which could fall into this category include *force majeure*, such as earthquake, flood, fire, landslip, pestilence and disease, and those risks associated with labour, plant, equipment and materials, technology and management. As increased project costs and time overruns are the main risks during the pre-commissioning period, it is essential that the contracting risks be clearly defined. Lenders to a project are particularly sensitive to the construction contract as such risks are principal causes of projects in trouble. The contract should be clear and preferably be on a fixed price turnkey basis with meaningful penalties for time overruns and performance failure. *Force majeure* and strike relief should be minimal. Risks in ground conditions should be identified and priced and preferably the obligation should be accepted by the contractor.

On-demand performance bonds need both to be set at levels commensurate with the level of risk, and capable of bearing such risk. An adequate percentage of retentions should similarly be insisted upon. Furthermore, the project management structure developed for the project should avoid variations and limit opportunities for design changes and extensions to the time for completion. Associated infrastucture risks are particularly important if connecting or approach roads have to be constructed by a specified date by the public sector. This problem may be exacerbated in the case of a cross-border project in having to deal with two governments and their departments. Risks associated with infrastructure can be reduced by fixing a programme for hand over, for example, of access roads, with compensation payable by a government in the event of a delay on its part.

Operational Risk

As well as the risk associated with the physical operation of a facility, such as plant falling into disrepair due to neglect or negligence, damage to equipment or part of the project asset, other operational risks can include default or insolvency, operation economics, training of operatives, complexity of operation and operational interruptions. Other major risks which can be categorised as operational are those associated with a feedstock contract regarding the raw materials to be used and the off-take contract. These risks are particularly prominent in international power projects where the fuel is imported and thus purchased in the international marketplace which is subject to global fluctuations.

Latent defects in the project are of particular importance in certain types of design, particularly designs of a novel nature, and which only become apparent during operation. Some projects have to undergo extensive repairs within the design life because of unforeseen circumstances. Latent defects in the project may be minimised by increased attention to monitoring of the design. Detailed design coupled with adequate financial contingencies made available at the outset of the project to cover potential latent defect problems can provide additional comfort to the project.

Financial Risk

This category arguably covers the broadest range of risks associated with major infrastructure projects. Financial risks include those associated with the mechanics of raising and delivering of finance, and the availability of working capital. They can also include foreign exchange risks and debt service risk, where limited or non-recourse project finance is employed. Such risks may arise during the operation phase when the facility is running to specification, but do not generate sufficient revenue to cover operating costs and debt service; and those associated with the take and/or pay terms, and the effect of escalation clauses over the operating period.

Financial risks can relate to the cost of servicing a loan, default by the lender, loan period, cash flow milestones, the type of and changes in the interest rate and currency mis-matches. Some projects are particularly sensitive to interest rate fluctuations: generally projects where the risk lies in its capitalisation being based on a forecast interest rate which is not borne out in actuality. Such risks increase the cost of financing a project. These risks can be tackled by complex financial instruments, such as index-linked bonds and long-dated swaps, and fixing or 'putting a cap on' the interest rate. The second Severn Crossing in the UK includes, for example, a twenty-one year index linked debt with tolls being indexed similarly.

The effect of a rise or decline in inflation can be particularly severe for a project if combined with changes in the interest rate. Both lenders and equity investors may in financing a project seek to protect themselves against this risk by incorporating price escalation clauses in the off-take agreement (in the case of power projects), or by making provision in the concession agreement to allow the promoter to increase tolls (in the case of a tolled road project). Where an infrastructure project sells its output or service to the local economy, it will receive its earnings in local currency: the risk is increased when different currencies are involved. Lenders and equity investors will need to be assured that they will be able to recoup their original investment, together with interest and dividends, and will include such matters as authorisation to convert local currency earnings into foreign currency. They will also seek to ensure that sufficient foreign currency is available in the country's banking system to facilitate the conversion and that this can be achieved at a favourable rate.

Revenue Risk

The risks associated with revenue generation are often considered on the basis of meeting demands. The accuracy of demand and growth data, the ability to meet increased demand and tariff/toll variation formulae are all classic revenue risks. Market-led revenues are far more uncertain than those based on pre-determined sales contracts, thus promoter organisations will often seek contract-led revenue streams. For example, in a toll road facility a promoter may approach haulage contractors to enter into take and/or pay arrangements for the use of the facility to reduce the risks associated with revenue generation.

Lenders often demand protection against factors beyond the control of the project's participants and which could affect the project's ability to generate earnings to service the debt. Such protection is often used to assure a minimum stream of project revenue. The most common forms of protection being arrangements whereby users, sponsors or third parties agree to make periodic payments in return for a given portion of output, services or use. The obligation to make payments is, generally speaking, unconditional, regardless of whether or not the product or service is delivered, as in take and/or pay contracts. Similarly, through-put agreements stipulate that pipeline users put a minimum amount of a product through the pipeline at periodic intervals and pay for the use of the pipeline irrespective of whether or not the stipulated amount of through-put is achieved. The revenue risk may be reasonably easy to ascertain in some projects, but extremely difficult in others. In the case of an estuarial crossing a captive market can provide reasonably accurate traffic predictions. However, where there is no captive traffic flow, as in the case of a new toll road, the issues relating to potential traffic flows at various levels of toll become extremely complex. Revenue risk could be reduced by obtaining more detailed forecasts, the use of sophisticated computer models and government agencies providing more information to bidders at the bidding stage.

In mitigating risk under this heading, potential for risk sharing between the public and private sector is available if economic benefits other than return on the investment are taken into account. The shared traffic risk between the principal and the user is possible by allowing increased tolls and increasing the concession period if the traffic falls below an agreed estimate for traffic flow. Another option is for a government to provide a revenue support guarantee if traffic forecasts fall below an agreed level. Other possibilities include the provision of subordinated debt arrangements by a government and maximising guaranteed revenue by way of user agreements, such as the arrangements with the British and French railways on the Channel Fixed Link project.

Political Risk

Political risks can be associated with both local political powers, such as the change in policies or parties and/or those risks generated by political entities outside of the jurisdiction. Political risks can be related to a government's attitude towards allowing profits on infrastructure projects, repatriation of profits and changes in regulations. Other risks which fall into this category include expropriation, nationalisation, changes in taxation, rationing of production and forced sale of the asset or its off-take. Before proceeding, potential parties to a project need to assess the enthusiasm of the host government and in particular whether there are any political difficulties preventing a project from going ahead: fiscal and policy changes constitute real threats to a project. A domestic regulatory change would not be considered a major issue if the concession agreement made provision for compensation in the event of such change. Another alternative is to offer a promoter protection by reducing future competition, such as Eurotunnel's option to build a second fixed link across the channel.

Legal Risks

Legal risks can be sub-divided into those that can be associated with the host country, and those that can be more particularly linked to the procurement arrangements. Falling into the former category are the nature of the existing legal framework, changes in laws and regulations during the life of the project, and conflicting economic community (if applicable), national and regional laws. Where projects are to be constructed across national boundaries, different laws and conflicts between legal systems can add to pre-construction delay risks. Into the latter sub-division can be placed the risks associated with the type of procurement or concession arrangements; changes in the obligations under the legal framework; and resolution of disputes. Such risks can also extend to the ability of the promoter legally to enforce the provisions of a concession agreement and requirements of statutory undertakers. Since the organisational structure often associated with major infrastructure projects involves numerous legal agreements between the organisations involved, and which must operate within the legal framework of the host country, the legislation affecting that structure can be a great risk. In overseas projects the legal system of the host country may require the use of local companies and nominated suppliers to ensure compatibility with existing or proposed facilities operated by the principal.

Commercial Risks

Risks affecting the market and revenue streams, and hence the commercial viability of a project, can, broadly speaking, be classed as commercial risks, examples of which can include changes in demand for the facility, escalation of costs of raw materials, consumer resistance to tolls, convertibility of revenue currencies and devaluation. Each project has its own inherent risks. Demand risks are not normally controllable on a road project and thus promoters should be allowed to extend the operation period if demand is less than that which is predicted. Further a take and/or pay contract, or other through-put arrangements, could be used to reduce the risk of off-take demand being reduced by the user. In order to minimise foreign exchange risks, one can arrange the finance of a project in the same mixture and proportion of currencies as those anticipated from the revenue stream.

Environmental Risks

Environmental issues are assuming an increasing importance on a global basis. The construction industry like many other industries faces increasing pressure from both public concern and government policy directed at achieving significant reductions in environmental degradation. Such issues are a major aspect in the planning and design of major energy, water or transport infrastructure projects, and increased awareness and involvement by pressure groups is particularly noticeable with such major construction projects. Promoters considering any project in the UK which requires approval by an Act of Parliament are now required to provide an Environmental Statement for consideration by a relevant or interested Select Committee. The nature and timing of environmental assessments also need to be considered.

The location of a project can make it particularly sensitive to environmental risks. Emissions do not recognise boundaries and perceived environmental risks in one country may have far reaching effects in another country. For example, effluent discharge into a river which crosses a number of borders may create pressure from the downstream country to stop production or generation and/or require major technical upgrading of the facility under the concession agreement. Existing environmental constraints and impending environmental changes may differ from country to country. As well as the risk of existing environmental constraints, the risk of an environmental catastrophe on a new facility may lead to new environmental legislation. This may then expose a financial risk, during the construction and/or operation phases, associated with meeting environmental requirements that are subsequently imposed.

Environmental concerns do not relate only to the construction of projects. Much of the opposition is directed at the knock on effects of a new structure or plant, in terms of the long term pollution risk from the operation of the project in question. Thus, even when the project has been constructed, the risk will not necessarily diminish Whilst the significance of existing environmental requirements may be assessed and determined at the project appraisal stage, the risks associated with the introduction of new environmental requirements brought about through any number of routes may not be capable of identification. Notwithstanding which, the impact of such requirements may affect the commercial viability of the project at any stage.

Response to Risks

Having identifed and analysed the risks associated with a project, those risks should be apportioned to each of the parties involved. It is often suggested that risks should be assumed by the party within whose control the risk most lies; each party insisting on reward commensurate with the nature and extent of the assumed risks. The response to a risk should take into account the level of impact, the resources available to determine and implement the response and the costs involved in alternative responses. In this regard, risks can be averted by avoiding the activity with which the risk is associated. For example, the risk of steel corrosion of pipework may be avoided by use of a proprietary treatment. The impact of a risk may be reduced by lowering the probability of occurrence or by diminishing the extent of the loss. The effects of risks may be transferred from one party to another; for example, to the principal in a concession arrangement, or by insurance. Where the cost of risk transfer exceeds the expected cost of the risk by more than the premium required to cover the risk, then risk retention would be a valid response.

Guarantees in off-take contracts can be used to transfer risk due to changes in market conditions from the project users, take and/or pay contracts guarantee the project a future stream of revenues. Lump sum or turnkey contracts can be used to transfer cost overrun and other construction risks to the contractor. Some construction and operational risks can be covered by performance guarantees, completion guarantees, warranties and operating guarantees. A structured concession agreement could provide the basis for the allocation of risk to the organisations responsible for finance, construction, operation and maintenance and, where applicable, to those responsible for supply and off-take.

Most governments wish to maximise rather than optimise the transfer of risk from the public to the private sector. Such a policy can deter bidders from privately financed infrastructure schemes: the private sector is normally concerned with the cost of the scheme, the risk and associated delay. To achieve a successful privately financed project, a government will have to ensure that there is no imbalance between the risk and return. If the cost of tendering for a private finance contract is too high bids will be discouraged. Governments need to consider what efficiency gains might result from the transfer of each type of risk to the private sector and how it would affect the risk/reward ratio for the promoter of privately financed infrastructure projects. As a general rule the government should be prepared to retain some or all of the risk where:

- it does not threaten the incentive for efficiency gains by the private sector;
- the risk is largely outside the control of the private sector and there is thus little to be gained from transfer to the private sector. Broadly speaking, construction and performance risk are controllable whereas demand and financial risk are at least to some extent not controllable;
- the risk can only be transferred at a cost to the private sector which is far higher than retaining the risk in the public sector.

However, the interdependence between the risks complicates this general rule. In particular, financial risk is largely outside the control of the private sector, but the assumption of this risk by the private sector will have a favourable incentive effect on project cost elements which are largely subject to controllable risk, such as construction and performance. The impact upon the project as a whole must therefore be carefully considered.

A number of measures could be introduced which would reduce the risk to the private sector while not reducing the incentives through efficiency gains:

- governments need to recognise that large projects attract high risks and therefore rewards must be sufficiently high to attract equity investors;
- specific tender documents and clear government requirements would in turn reduce the costs of bidding and simplify the procedure for the evaluation of bids by the public sector;
- financing risks could be reduced by providing a range of BOT projects so that investors and promoters can spread the risk across a portfolio of investments;
- a government could share the financing risk in the project by subordinating debt, bearing part of the capital cost or taking an equity stake in the project;
- demand risk may be shared between the government and the private sector through a variable concession period. If demand is less than expected, the contract period can be extended to allow the private sector further time to recoup its investment. As this may weaken the incentive for the private sector to influence demand through improved service quality (performance risk), it should be restricted to projects where demand risk lies largely outside the control of the private sector; and
- a major risk to some projects is competition from existing projects. For example, a major risk to privately financed toll roads is an untolled road network. In this situation it may therefore be appropriate to consider tolling competing roads.

Conclusion

Participants in a project must fully appreciate the nature and extent of the risks they assume. Participants, promoters and lenders alike must remember that an unfair or disproportionate allocation of risk to participants, who cannot withstand the impact of its occurrence, can be damaging not only to that participant, but also to the project as a whole. Measures can be taken in mitigation, but rarely will a risk be fully extinguished. The risks inherent in privately financed infrastructure projects are greater than those associated with traditional forms of contract since the revenue generated by the operational facility must be sufficient to pay for the construction, operation and maintenance, and finance. The uncertainty of demand (and hence revenue), cost of finance, length of concession periods, effects of commercial, political, legal and environmental factors are but a small number of the risks to be considered by promoter organisations.

There is little doubt that need, be it from the public or private sector, is the principal factor driving the procurement of a major infrastructure project. Many environmental benefits can be demonstrated from investment in infrastructure; construction of alternative or additional forms of transportation to reduce congestion on existing facilities; investment in more fuel efficient and environmentally friendly power stations and water treatment facilities, which will bring more local benefits to an area concerning noise levels, and air and water quality.

Investments in improvements to existing infrastructure, particularly transportation, and other facilities largely used by the general public, can reduce the risk of injury or death which in turn can reduce the burden on health care facilities. The construction of infrastructure and industrial projects often involves the direct increase in local employment and will often continue throughout the operation of the facility: an increased demand for local equipment and materials is manifestly to the economic benefit of the country associated with the investment. A local authority demonstrating a commitment to energy efficiency and protection of the environment may be able to attract inward investment into the local economy by virtue of the positive local environrnent created, which in turn may extend to improved standards of health as well as reduced operating costs.

Successful infrastructure project development has as two of its most crucial ingredients the political will of the host government to 'champion' the project and, so far as is possible, to offer political stability throughout its operational life, and the host government's understanding of promoters' and lenders' risk/reward sharing expectations. Support in a variety of forms, such as the legal and administrative environment, convertibility of revenue earned, logistical measures, needs to be forthcoming from the host government to minimise the risk exposure of both promoters and lenders. An imbalance of the risk/reward ratio will only serve to discourage both private investment and more particularly the entrepreneurship and initiative of private sector promoters.

15. Major Infrastructure Projects and their Environmental Impact II *

Joanna Higgins

Synopsis

This paper provides an overview of the areas of law and consequent financial risk which promoters, funders and their advisers must take into account in the proper management of a project. The paper deals with risks in the order in which they are likely to be faced by the project team, commencing with planning consent.

Introduction

Any major infrastructure project will impact upon the environment in which it is constructed and subsequently operated. This paper takes as its perspective the way in which environmental impact is increasingly a factor which project promoters need to take into account when assessing the viability of their schemes. The paper analyses the risks associated with environmental impact and indicates the increasing likelihood of projects constructed in the UK suffering delays and/or unexpected costs as a result of environmental legislation. The perspective taken here is that of the privately funded project promoter.

---

*   This paper is based on the author's MSc dissertation which was awarded the Society of Construction Law Prize for 1994.

It is obviously important that one should define what is meant by the "environment". A description of what is to be considered as the environment is to be found in Annex III to EC Directive 85/337 where aspects of the environment likely to be affected by a relevant project are to include population, fauna, flora, soil, water, air, climatic factors, material assets, including the architectural and archaeological heritage, landscape and the inter-relation between these factors. The definition is reiterated in the implementing English regulations which is the Town and Country Planning (Assessment of Environmental Effects) Regulations 1988. These are more wide-ranging definitions than the provision of the main English statute on the environment, the Environmental Protection Act 1990, which defines the "environment" in s1(2) as "all or any, of the following media, namely, the air, water and land ....". There are inherent difficulties therefore in assessing what the impact of a project on the environment may be, because of the uncertainties of precisely what the environment can be said to be.

This paper takes as its premise that the impact of an infrastructure project will be both during the construction phase and potentially also once construction is completed, through the operation of the finished product itself. Impact on the environment can be through the noise, dust and vibration routinely caused by the construction process. It can also be through the escape of substances from the construction site into watercourses or other land. It also comprehends the impact on the environment in the wider sense in terms of protected areas of countryside, through which, for example, roads may be constructed. This paper offers an overview of the risks assumed by the project promoter of infrastructure projects, in terms of the possible time and cost implications of environmental issues. It deals with the importance of environmental issues at the application for planning permission stage. It then deals with the construction process but limits its terms of reference to those factors likely to be common to all construction processes, namely noise, dust and vibration. It then goes on to deal shortly with the impact of the constructed project with particular reference to the remedy of judicial review.

This paper acknowledges that there is increasingly an awareness by the public of environmental issues and ways of using the judicial process and other means as a way of dealing with these issues. This involvement can be characterised as "active" or "passive". The former are those who take action in relation to projects, such as the M11 Link Road protesters whose well-publicised attempts to protect the environment have recently come to an end. The latter are those who are themselves affected by the environmental impact from the project and seek compensation in relation to that harm suffered by them rather than to prevent the works.

The effect of this public awareness is potentially to delay projects, on the one hand by use of physical intervention, and on the other to increase costs by incurring a liability in damages arising out of actions in nuisance and/or negligence perhaps long after the project has been completed and in respect of hundreds of local residents affected by the works. This is increasingly likely to have to play a part in project promoters' analyses in the future in view of the number of infrastructure projects which are taking place or to take place in populated areas. It should also be borne in mind by project promoters that an associated risk for project promoters arising out of environmental issues is the adverse publicity and effect on share price that can arise. The increasing array of remedies available to both of these categories of complainants is recognised by this paper; the increasing use of the remedy of judicial review in the environmental sphere is one example. Another recent example is the increased liability for property blight on the part of the Department of Transport.[1]

---

[1] *R. v. Secretary of State for Transport, ex parte Owen and Another* Court of Appeal (Civil Division) Transcript, 30 June 1994

The Planning Process

The failure to obtain permission to construct a project is an obvious risk. Any project promoter will seek to obtain planning permission at an early stage. Similarly, if a public inquiry is thought likely to be necessary, he will invest time and money in the preparation of evidence to substantiate the need for the project. These are likely to be commonplace occurrences and the risk of the project failing at this first hurdle will have been carefully analysed by the promoters' funders. What is a less well recognised and in many ways a less quantifiable risk is that of an environmental assessment being ordered pursuant to EC Directive 85/337.

On 27 June 1985 European Directive 85/337 was enacted. This required governments to ensure that environmental assessments were undertaken for certain public and private projects likely to have a major effect on the environment. This Directive was subsequently enacted into UK law by the Town & Country Planning (Assessment of Environmental Effects) Regulations 1988. This in turn is the subject of guidance in The Department of Environment Circular 15/88. Directive 85/337 *"on the assessment of the effects of certain public and private projects on the environment"* requires that the regulatory authority in the member state has to take into account the environmental impact of that development before a zoning permit is issued in certain categories of project. The underlying policy of the directive is "in preventing the creation of pollution or nuisances at source, rather than subsequently trying to counteract their effects." The environmental assessment aims to identify, describe and assess the possible direct and indirect effects of the project on the environmental factors referred to in the introduction above.

The Directive and the Regulations contain two lists of projects, those in Annex I (Schedule 1) where the regulatory authority must ask for an environmental impact assessment and those in Annex II (Schedule 2) which gives member states the discretion to act in cases where they *"consider that the characteristics of the project so require"*. Member states can establish criteria and/or thresholds necessary to determine which of the projects of the classes are to be subject to an assessment. A proposal published in the UK in March 1994 would tighten criteria for Annex II projects and add some new categories. The effect of the Directive and its implementing legislation is at its simplest level to add a further cost for the project promoter in compiling the necessary information to enable him to satisfy the local planning authority and/or the Secretary of State that the project should be granted planning permission. In relation to infrastructure projects, a number of these will fall within the terms of Annex I. These developments are therefore necessarily subject to preparation of environmental assessments. Other projects will fall within the provisions of Annex II. The risk for the project promoter here lies in the uncertainty of the need for an assessment to be undertaken.

The structure of the legislation provides for example that even if the local planning authority determines that an environmental assessment is not necessary, the Secretary of State may nonetheless call for one. There is guidance in the DoE circular referred to above, but this is itself vague, referring to "major projects of more than local importance", projects in "particularly sensitive or vulnerable locations" and/or having "unusually complex and potentially adverse environmental effects". In the author's view the risk averse project promoter will avert the potential delay associated with having an environmental assessment considered or imposed by the Secretary of State by obtaining an environmental assessment in any event, notwithstanding the further cost burden that this will impose on the project. A further factor militating in favour of this "risk avoidance" approach is the availability of the remedy of judicial review in respect of the planning authority's decision to grant planning permission with an environmental assessment.

An application can be made by affected third parties who can show a *prima facie* case that an assessment should have been made. Such an application will frequently be accompanied by an application for an injunction to preserve the status quo pending determination of the judicial review application. Although the application for the injunction will be judged on the balance of convenience in the normal way, it is a powerful argument that if the injunction preventing the works is not granted that the plaintiffs will be left without recourse. A further risk to the project promoter in this regard is the time span of judicial review applications which due to the onerous caseload of the courts, even on an expedited basis, could represent a delay to the project of 9 months. This can obviously have very expensive consequences for the project promoter. This paper will deal in more detail with the implications of judicial review as a project risk in the section of the paper dealing with the operation of the constructed project.

Liability for Environmental Impact during the Course of the Works

As far as the project promoter is concerned his exposure during the course of the works will fall under a number of main heads, namely

- Private nuisance
- Public nuisance
- Negligence.
- Statutory liability
- Strict liability under *Rylands v. Fletcher*[2]

Private Nuisance / Public Nuisance

Clerk & Lindsell on Torts[3] state that:-

---

[2]   (1868) LR 3 HL 330, HL, affirming (1866) LR 1 Exch 205.
[3]   Clerk & Lindsell on Torts, Sixteenth Edition, Sweet & Maxwell, London, 1989.

> *"In private nuisance ... the conduct of the defendant which results in the nuisance is of itself not necessarily or usually unlawful. A private nuisance may be and usually is caused by a person doing on his own land something which he is lawfully entitled to do. His conduct only becomes a nuisance when the consequences of his acts are not confined to his own land but extend to the land of his neighbour by:-*
>
> *(1) causing an encroachment on his neighbour's land...*
>
> *(2) causing physical damage to his neighbour's land or building or works or vegetation upon it...; or*
>
> *(3) unduly interfering with his neighbour in the comfortable and convenient enjoyment of his land.*
>
> *It may be a nuisance when a person does something on his own property which interferes with his neighbour's ability to enjoy his property by putting it to profitable use. It is also a nuisance to interfere with some easement or profit or other right used or enjoyed with his neighbour's land. (Nuisances of this kind, causing an interference with the enjoyment of land, are e.g. creating stenches by the carrying on of an offensive manufacture or otherwise, causing smoke or noxious fumes to pass onto the plaintiff's property, raising clouds of coal dust, making unreasonable noises, or vibration,...)".*

In relation to the burden of proof to be discharged in connection with the first two types of nuisance, nuisance is established by proving the encroachment or the damage to the land as the case may be. The situation of the land affected, the character of the neighbourhood and the surrounding circumstances are not matters to be taken into consideration. For the project promoter, by causing noise, dust and vibration in the construction of his works, he will fall within the third limb where the legal test of whether a nuisance has been caused is rather different:

> "...In nuisance of the third kind the personal inconvenience and interference with one's enjoyment, one's quiet, one's personal freedom, anything that discomposes or injuriously affects the senses or the nerves, there is no absolute standard to be applied. It is always a question of degree whether the interference with comfort or convenience is sufficiently serious to constitute a nuisance...".[4]

What is, however, clear is that the inconvenience must "*materially interfere with the ordinary comfort, physically, of human existence in the plaintiff's premises according to plain and simple and sober notions among English people*".[5] The project promoter does not therefore have to take his plaintiffs as he finds them, as with negligence actions, but is entitled to defend himself against private nuisance actions on the basis that the plaintiffs concerned were unusually sensitive. It is probable however that environmental impact claims based on construction work will be brought in both private and public nuisance.

"A person is guilty of a public nuisance ... who (a) does an act not warranted by law, or (b) omits to discharge a legal duty, if the effect of the act or omission is to endanger the life, health, property, morals, or comfort of the public, or to obstruct the public in the exercise or enjoyment of rights common to all Her Majesty's subjects ..."[6] Actions in public nuisance may be brought by the Attorney-General on behalf of a section of the community, by the relevant local authority under the provision of Section 222 of the Local Government Act 1972 or by an individual citizen if he can

---

[4] *Ibid* pp 1358-9
[5] *Emms v. Polya* (1973) 227 EG 1659
[6] Archbold's Criminal Pleading, Evidence and Practice, 44th Edition, Sweet & Maxwell, London, 1994, Vol. 2, §§ 31-50; *Attorney General v PYA Quarries Ltd.* [1957] 2QB 169, *per* Romer L J at 184

show special damage over and above that suffered by the community at large. In practical terms, because of the limitations of public funds, it is in fact probable that any public nuisance actions will in fact be brought by individuals, perhaps assisted by pressure groups.

Private nuisance actions have rarely been before the Court, but the impact for a project promoter of a claim in nuisance being brought can be great, whether that action is commenced during the course of a project or when it has been completed. The significance of a private nuisance action during the course of construction of the project is that the remedy of an injunction is available. Furthermore, fault on the part of the defendant does not need to be established to the same extent as in a negligence action and the fact that a defendant has used "best practicable means" is not an absolute defence in a private nuisance action as it is in most statutory nuisance actions. The Courts operate a balancing exercise between the defendant's right to conduct his operations and the plaintiff's right to have the use and enjoyment of his neighbouring property without undue interference.

It is well accepted law that the locality of the project is a significant factor in determining whether a remedy will be granted. This has particular significance for infrastructure projects being undertaken in the UK in close proximity to residential areas. It is difficult, however, for the promoter to analyse the extent of the risk undertaken because of the fluidity of the legal standard against which he must measure his project. In *Colls v. Home & Colonial Stores Limited*,[7] it was said that:

> "*A dweller in towns cannot expect to have as pure air, as free from smoke, smell, and noise as if he lived in the country and distant from other dwellings, yet an excess of smoke, smell and noise may give a cause of action, in each*

---

[7] [1904] AC 179

of such cases it becomes a question of degree, and whether in each case it amounts to a nuisance which will give a right of action". [8]

A recent decision of some significance in this area is the well-known case of *Gillingham v. Medway (Chatham) Dock Company.* [9] There, the trial Judge ruled that the effect of grant of the planning permission was to change the character of the neighbourhood to the extent that a nuisance action will only be successful if it can be shown that the nuisance caused exceeds that which was necessarily comprehended in the grant of planning permission for the development. There has been some considerable disagreement about the importance with which this case will be invested in the future but it would certainly be wise for a project promoter to assume that there is still a residual risk that an action in private nuisance could still lie, even if the planning permission for his project appears to comprehend the actions which he has carried out.

It has been said that there are special allowances made by the Courts for construction operations. Indeed, the Courts are inclined to take the view that construction projects inevitably generate noise and dust and neighbours must therefore tolerate a certain degree of disruption:

> "... when one is dealing with temporary operations, such as demolition and re-building, everybody has to put up with a certain amount of discomfort, because operations of that kind cannot be carried on at all without a certain amount of dust. Therefore, the rule with regard to interference must be read subject to this qualification, and there can be no dispute about it, but in respect of operations of this character, such as demolition and building if they are reasonably carried on and all proper and reasonable steps

---

[8]   *Ibid, per* Earl of Halsbury LC at 185
[9]   [1992] 3 WLR 449

are taken to ensure that no undue inconvenience is caused to neighbours, whether from noise, dust or other reasons the neighbours must put up with it".[10]

Pugh and Day[11] are of the opinion that:

"There is however one important distinction between establishing liability in private and public nuisance actions. It is a pre-requisite of establishing liability in private nuisance that the plaintiff has an interest in affected land. This is not a precondition of a public nuisance action. For most people the place where they are environmentally disturbed and inconvenienced is in their home and, someone in the home will in most cases have legal interest in it either as an owner or tenant. This is sufficient for that person to bring a private nuisance action. However, in an instance where a whole family is affected an action in private nuisance could only be brought by a person with an interest in land, so that other adults and children could not bring a private nuisance action".[12]

It is of interest to note that, in considering injunctions based upon the balance of convenience, the Courts are prepared to take into account the financial implications of an injunction to the project promoter. In the above case, the Court of Appeal substantially set aside an injunction on the grounds that the restricted hours would give rise to a cost overrun to the order of £1.2 million. However, in *Hart v. Agha Khan Foundation*[13] case, the Court of Appeal made it clear that the standard of precautions expected of a project promoter will also be affected by the probable duration of the project. For

---

[10] *Andreae v. Selfridge & Co* [1938] Ch 1, *per*, Sir Wilfred Greene MR at 5 - 6.

[11] Pugh, C. & Day, M., *Toxic Torts*, Cameron May, London, 1992.

[12] *Ibid* p.110

[13] Unreported, Court of Appeal 13 February 1981; Bar Library Transcript 1980M 4893.

major infrastructure projects, it is therefore likely that state of the art techniques to minimise dust, vibration and noise will be required.

The project promoter should also be aware that another line of authority may provide him with a defence, namely that of statutory authority provided that his project has been authorised by statute. The line of cases from *Allen v. Gulf Oil Refining Company*[14] may provide a total defence to a private nuisance claim provided that the promoter's activities fall within the ambit of the authorising legislation. The risk to the project promoter lies therefore in the uncertainties of whether the construction of the project will be determined by a court to have overstepped the boundaries of what is acceptable disruption for local residents.

Negligence

Negligence actions have a relatively limited role in risk assessment for the project promoter since injunctions are not available, pure economic loss is not recoverable and exemplary and aggravated damages would appear not to be available.[15] From the promoter's point of view, a further significant advantage is that defendants are not generally liable for the negligent acts of their independent contractors providing that an apparently competent contractor has been selected. There is, however, the advantage to plaintiffs' lawyers that compensation for personal injuries is available and the plaintiff need have no interest in land to claim. In practical terms, however, since fault must be established, for most project promoters the risk associated with a successful action being brought in negligence against them is relatively low.

---

[14] [1981] 2 WLR 188
[15] See for example *AB v. SW Water Services Limited* The Times, May 8, 1992: The Camelford Water Litigation.

Statutory Liability

Particularly noteworthy as far as project promoters are concerned are the following:

Control of Pollution Act, 1974, Sections 60 and 61
Environmental Protection Act, 1990, Sections 79 to 81.

The provisions of the Control of Pollution Act 1974 are relatively well-known. Section 60 provides for a local authority to serve a notice imposing requirements as to the way in which works are to be carried out in order to control noise on construction sites. The "works" are defined in s.60(1) and include "any works of engineering construction" as well as certain other forms of works. The power of the local authority is wide-ranging in what it may specify in the notice but can include specifying working hours for the site or specific plant to be used (or not be used). There is provision for appeal against the terms of a notice to the Magistrates Court. There is however some question as to how effective such an appeal is in practical terms since Magistrates are ill-equipped generally to deal with the technicalities of the acoustic impact of engineering plant on site. The prudent promoter will commonly therefore take advantage of the provisions of s.61 of the Control of Pollution Act which allows "a person who intends to carry out works" to apply for a prior consent for works. The application must provide particulars of the works and the method by which they are to be carried out and the steps proposed to be taken to minimise noise resulting from the works.

It is fundamental in analysing risk connected with the project that the promoter takes into account the impact which failure adequately to deal with noise impact could have on the project. Any person who knowingly carries out the works, or permits the works to be carried out in contravention of the terms of a s.61 consent is guilty of a criminal offence. It is conventional for such consents to require the consent holder to ensure that best practicable means ("b.p.m.") are employed to reduce noise to a minimum, reflecting the local authority's obligation under s.60 to have regard to ensure b.p.m. B.p.m. is defined in s.72 of the Act.

The obligation to use b.p.m. will be expressed to reduce noise to a minimum. Although s.72 provides for certain factors to be taken into account, promoters need to be aware that the obligation is not merely to reduce noise. S.72(2) of the Act states that "practicable" means reasonably practicable having regard among other things to local conditions and circumstances to the current state of technical knowledge and to the financial implications. For any major project it is unlikely in practice however that financial implications will be taken into account by local authorities. Promoters should also be aware that fines for contravention apply to each such contravention i.e. each piece of machinery on site failing to exhibit b.p.m. on any particular day the promoter can be subject to a fine. Furthermore, an offence is committed even if the noise limits are not broken where plant fails to exhibit b.p.m.

In view of the above, the promoter may wish to avoid obtaining s.61 consent. He will wish therefore to ensure that the contractor does so rather than allowing the works to take place subject to the risk that a s.60 notice may be served. The promoter needs carefully also to consider the contractual framework between himself and the contractor and any consultants employed to ensure that environmental obligations are clearly apportioned. Promoters should note that if they and/or their contractor ignore the service of s.60 notices or s.61 consents, the Courts are prepared to grant injunctions in support of such notices or consents.[16]

The Environmental Health Act 1990 was enacted to consolidate existing environmental legislation and forms the major source of environmental legislation in the UK. As far as the project promoter is concerned, however, the Act has only limited relevance in the

---

[16] *City of London Corporation v. Bovis Construction Ltd.* (1987) 40 BLR 7.

areas of noise, vibration and dust which are dealt with by this paper. Section 79 of the Act creates statutory offences relating to noise (including vibration) and dust being prejudicial to health or a nuisance. The project promoter needs to be aware that best practicable means will again operate as a defence and should form part of the risk management strategy of the promoter. This can best be achieved by ensuring that the contractor executing the project is contractually bound to use best practicable means to reduce noise, vibration and dust pollution. It is also prudent to ensure that contractors exercise appropriate control over their sub-contractors. The primary risk in relation to the Environmental Protection Act is that an abatement notice may be served on the contractor (s.80) thereby affecting progress of the works, if hours of working are limited, for example. Since breaches of abatement notices are also criminal offences, punishable by maximum fines of up to £20,000 per offence, the promoter should also seek to avoid breaches by the contractor in order to avoid adverse publicity for the project. It is, however, a defence to a prosecution under this Act, that a s.60 or s.61 Control of Pollution Act notice or consent is in force and is being complied with. It should be noted that such an argument would not offer an automatic defence in a common law nuisance action.

Strict liability in Rylands v. Fletcher

A defendant will be liable under the rule in *Rylands v. Fletcher* if certain criteria are fulfilled. He must bring matter on to his land, in circumstances that make what he is doing a special or "non-natural" use of the land. The matter must then escape from the land. Finally, as a natural result of that escape, the matter must cause harm to the plaintiff. If these criteria are established, the defendant will be absolutely liable even if he took all possible precautions to prevent damage.

In *Cambridge Water Company v. Eastern Counties Leather plc*[17], Lord Goff considered the historical development of the *Rylands v. Fletcher* doctrine. He treated the case as a sub-category of the main body of the law of nuisance. He also looked at the American principle of liability for ultra-hazardous operations and considered that this principle was not applicable in the law of England and Wales. Strict liability for environmental harm, if imposed at all, should be imposed by Parliament rather than Courts. The main thrust of Lord Goff's judgment was that in both nuisance and in *Rylands v. Fletcher*, liability had to be based on foreseeability of harm. The judgment, however, found the Cambridge Water Company liable and it has been widely predicted that the doctrine in *Rylands v. Fletcher* will have an increasing part to play in the future. This obviously may have major financial implications for project promoters, since liability is absolute if the criteria are met. It is no longer clear what the court will take non-natural user of land to mean and it is difficult for promoters as their advisers to quantify their exposure in this regard.

Class Actions

A new phenomenon in the context of construction and infrastructure projects is the advent of Class Actions brought by those affected by projects. Class Actions have previously been associated with primarily product liability cases, most prominently in the pharmaceuticals and tobacco industries. There have in recent years, however, been a number of instances of legal aid being granted and proceedings pursued in respect of claims arising out of construction work. The most prominent of these relate to the regeneration of London's Docklands and in particular the construction of the Canary Wharf Tower and the Limehouse Link cut and cover road.

---

[17] [1994] 1 All ER 53

In respect of the latter judgment on various preliminary issues has been handed down and is now subject to appeal on various grounds and this will have an impact on project promoters in the UK in the future.[18]

In common with many recently promoted private infrastructure projects such as the Jubilee Line extension and the Heathrow express link, the Limehouse Link road was constructed in close proximity to residential areas. In that case the action is based on allegedly excessive amounts of dust generated by the construction works. The pleaded causes of action are in negligence and nuisance and over 600 plaintiffs were granted legal aid to bring the proceedings. The Judge found, for example, that annoyance and discomfort alone without further damage cannot found an action in negligence. He went on to find however that "damage", i.e. damage sufficient to enable the Plaintiff to bring a claim in negligence, was sufficiently met by the deposition of dust on land in sufficient quantities to impair the usefulness of that land by, for example, requiring curtains to be repeatedly cleaned.

One of the consequences of these developments of the law is that the project promoter must ensure that those undertaking the project will do so in a way which minimises the prospect of these forms of action being brought. This obviously impacts upon the terms of appointment of any consultants appointed and the contractor or joint venture or other entity undertaking the works. It also impacts upon the insurance position of those parties. Promoters must be aware that this is a developing area of the law and it is undoubtedly the case that similar actions will be brought in the UK in the future. It is also the case that even if individual recovery may be small, with large numbers of plaintiffs the financial impact on the project may be large. The legal costs of defending such claims may also constitute a

---

[18] See *Hunter & others v. London Docklands Development Committee*, unreported.

significant overhead or result in an increase in insurance premiums in the future.

Liability for Operating Plant and Judicial Review

It is trite to state that many infrastructure projects will impact upon the environment once constructed. This will range from emissions from cars on newly-constructed highways to emissions from power stations. Many of these areas are the subject of specific legislative regimes. Others will potentially be governed by the common laws in the normal way. A less obvious risk is that associated with the intervention of affected parties through the judicial review process, seeking not compensation but to challenge, for example, the right to operate the plant. Obviously this can have wide-ranging consequences for projects and is therefore the focus of this part of the paper in relation to operational plant.

It is important to note that judicial review of governmental or quasi-governmental bodies is increasingly a remedy which environmental lobbyists are seeking to use and they are meeting with increasing success. The judicial review process can be used to *"police"* both at the stage of the grant of planning permission and subsequently in relation to, for example, decisions taken to operate the facility where a further governmental permission is required. It is also significant to note that the Courts are increasingly accepting that environmental bodies such as Greenpeace have a valid role in ensuring that legitimate environmental issues are fully considered by government bodies.

This of course has the effect that the project promoter in the private sector may find himself embroiled in judicial review proceedings in respect of governmental or quasi-governmental decisions which relate to the environmental impact of his project. An informative case on the development of the law in this area is *R. v. Inspectorate*

of Pollution and Another, ex parte Greenpeace Limited (No.2).[19]
The case arose in connection with the British Nuclear Fuels Limited
("BNFL") Thermal Oxide Re-Processing Plant ("THORP") at
Sellafield. This case was one of a series of cases brought by
Greenpeace in connection with THORP. The first question which
the court was considering in this case was whether Greenpeace had
locus standi to mount judicial review proceedings in respect of this
decision. The factual background of the case related to the grant of
new authorisations and variations of existing authorisations to enable
BNFL to test the new plant before it became fully operational.
Greenpeace's application was to quash the Inspectorate of
Pollution's decision to vary the existing authorisations and an
injunction to stay the implementation of the varied authorisations.
Greenpeace's application was dismissed by the judge but the primary
interest of the case as a development of the existing law relates to
Otton J's detailed discussion of Greenpeace's *locus standi*.

BNFL had argued that Greenpeace had no standing to bring judicial
review proceedings on the basis that they had no legitimate interest
in the decision in question and that therefore the grant of leave
should be set aside and the application disallowed regardless of
whatever the judge's views were on the merits of the case.

The Judge dismissed the view which had formerly held sway that
such a body had no legitimate interest in bringing judicial review
proceedings because they were in the vocabulary of the law relating
to judicial review a *"busybody"*. Otton J analysed the position and
held that the question of Greenpeace's interest arose at the stage of
the ultimate decision on whether judicial review should be granted
rather than at the application for leave stage. In this he stated that he
was following the House of Lord's decision in *IRC v. The National
Federation of Self Employed and Small Businesses Limited.*[20] Otton
J took the matter further in a long section of his judgment and in

[19] [1994] 4 All ER 329
[20] [1982] AC 617; [1981] 2 WLR 722.

effect sanctioned the status of such bodies as a means of protecting the rights of individuals. The Judge said that it was advantageous both to the Courts and to individuals affected that a well informed challenge be mounted which would also assist BNFL who would be left with a remedy in costs which they might otherwise not have had if the proceedings had been brought by individuals with the benefit of legal aid. The Judge did however note that it must not be assumed that Greenpeace or any other interest group would automatically be afforded standing in any subsequent application for judicial review in whatever field it might have an interest. This would be a matter to be considered on a case by case basis.

What is, however, of considerable interest in the case is that it is increasingly likely that bodies such as Greenpeace will bring judicial review proceedings in respect of a project having an environmental impact and where it can be said that the authorising body has not taken sufficient regard of environmental issues. This contrasts with the position previously which is that judicial review proceedings were primarily a remedy available to aggrieved individuals. This meant in practical terms that unless those individuals were wealthy or, alternatively, were able to bring the proceedings funded by the legal aid system, many decisions which might now perhaps be challenged would not have been because of the limited resources of the litigants. This is particularly relevant in the field of environmental impact where individuals would be unlikely to have access to scientific data which might well be necessary in order to substantiate their application. It is likely therefore that judicial review proceedings will increasingly form a risk for project promoters.

A further element of risk for the project promoter, primarily but not necessarily associated with the potential remedy of judicial review is the possibility that the project may be the subject of an Article 177 Reference (pursuant to the EC Treaty) to the European Court. Such as reference is in order to seek guidance from the EC Court on EC law and/or the implementation of that law in the UK. Such a reference might be made, for example, when the basis of the judicial review application is that the UK government's decision breaches relevant EC legislation. In that case, at any stage in the judicial review proceedings, the UK court of its own volition or at the request of the parties may refer the issue to the European Court for a determination. Once the European Court's decision has been made it will then be remitted back to the UK court. This can cause major delays to any project particularly if an injunction is granted in the interim preventing commencement of construction or operation of the plant.

Conclusion

First and foremost, the project promoter needs to be aware of the environmental issues relevant to his project. This paper deals only with certain limited forms of risk that a promoter has to have in mind. For example, the promoter may suffer major financial or temporal setbacks as a result of the construction operations polluting watercourses in its vicinity. In order to manage these risks, the promoter must consider at the outset and before problems arise which parties to the project or third parties, such as insurers should take responsibility for them. A number of insurers are now seeking to restrict the extent to which they indemnify their insured in respect of environmental liability, and this is a factor which may need to be taken into account. He must consider how the contractual arrangements between the various parties to the project should reflect the environmental risks. He must also be aware that environmental law is a rapidly developing area and one with which he and his advisers must fully acquaint themselves. Finally, the promoter must accept that certain risks to the project fall outside his control but may have fundamental effects upon the project. He must therefore analyse the probability of those risks occurring against their possible consequences with a view to satisfying himself and the project financiers of the viability of the project.

Part IV

Choice of Contract

16. Structuring Contracts for the Achievement of Effective Management [*]

John Perry

Synopsis

This paper describes the genesis of the New Engineering Contract, in terms of planning and objectives, and discusses continuing research into contract structure and drafting and the effect on performance and the achievement of objectives.

Introduction

In the Inaugural Michael Brown Foundation Lecture,[1] Professor Hugh Beale developed the theme of a multidisciplinary approach to Construction Law. He showed that to achieve a deeper understanding of construction contracts and improvements in the way they work needs a multidisciplinary approach, not just between engineers and lawyers, but also the involvement of other disciplines such as economics and social science. He addressed the difficulty of deciding a syllabus at postgraduate level for students who have already specialised in law or engineering. Those students now need greater depth in the topic of

---

[*] This paper was originally delivered to the Michael Brown Foundation, Centre of Construction Law, King's College London, May 16, 1994 and is reproduced by kind permission of Professor Perry.

[1] Beale, H. *A Multi-Discplinary Approach to Contruction Law.* The Michael Brown Foundation Inaugural Lecture. Kings College London. March 1992.

construction law on the one hand and breadth in terms of appropriate knowledge of new disciplines on the other. He also stressed the importance of research to both teaching and the development of the work at the Centre.

I want to try to take forward the theme of Research and Development and to relate this to the development of new forms of contract. There are several reasons why I believe the twin issues of R & D and contract forms are important. One reason concerns the need of the industry and clients which we serve. To Research and Develop new procedures for contract management and new approaches to contracts will benefit the profession by benefitting their clients. Civil Engineering, Project Management and Law are all service industries. We exist only because there are clients for our services. I have long held the view that the main purpose of academic activities, and especially R&D, in these subjects must be to help serve clients and society better.

Another reason is that after twenty years of teaching and researching in project management, including work on various contractual arrangements, I am amazed at how much there is still to learn and understand about the workings of construction contracts. In the area of construction contracts change occurred slowly and almost imperceptibly over a period of many decades. Such change that did occur was in the nature of fine tuning but we have recently witnessed both rapid and radical change including new forms of contract, new forms of dispute resolution and increasing use of contractual arrangements, such as design and build, which modify dramatically the conventional relationships between the parties. R&D can both lead the way to further changes and, by providing an accumulated body of knowledge, aid decision makers in making their choices between the options now available.

New Engineering Contract

The title of my lecture 'Structuring Contracts for the Achievement of Effective Management' reflects my particular interest in the role that contract forms can play in achieving good management of construction projects. I hope it will also show the vital role that the professionals who actually do the work can play and must play in R&D on this topic. The title was intended to raise questions. What does 'structuring contracts' mean. It was intended to refer to forms of contract not to contractual arrangements such as Design and Build or Management contracting, although there is an interaction there. To a civil engineer structure means form and shape - the Grand Plan, and is also concerned with the individual elements - the foundations, columns, beams and bracing, which we try to integrate in an optimum way to carry the loads imposed. In a similar way the structure of a contract form needs a Grand Plan which can be expressed in terms of objectives.

The Grand Plan for New Engineering Contract[2] was expressed very simply in terms of Flexibility, Clarity and Good Management. These were refined into sets of detailed aims and principles which were designed to meet the demands placed on a contract and the problems likely to be faced by it in the most efficient way. For a complete picture you will need to refer to NEC itself but I have listed some of the design principles which, combined with other elements, made up the complete structure of NEC. This is what I mean by structuring contracts:

- a choice of contract strategy;
- clauses divided between core,main options and secondary options;

---

[2] The Institution of Civil Engineers. *The New Engineering Contract.* First Edition. Thomas Telford Services Ltd. London 1993.

- a single procedure for compensating the Contractor for change and risk;
- a speedy procedure for dispute resolution;
- separate roles for project manager, designer, supervisor and 'disputes assessor';
- tight time periods for actions, replies and decisions;
- specific rules for approval/disapproval;
- advance quotations from the Contractor for proposed changes;
- compensation for change and risk based on the effect on actual cost and current programme;
- programme to include method and resources statements.

The title might have been improved by using the phrase 'Designing Contracts'. A good design has clear objectives and is based upon sound design principles. It is rare to find that there is a unique solution which makes the design fit for its purpose and options have to be considered at both conceptual and detailed levels. It is common to find that a particular option is good at meeting certain requirements but weaker at meeting others and compromises have to be made. This was my experience in designing NEC and I would be surprised if improvements cannot be made in subsequent editions. I have not specifically mentioned the drafting of contract terms. I have no wish to denigrate the importance of careful drafting, indeed its importance has been reinforced by my experience with NEC. What I have tried to do is to emphasise the importance of overall structure when developing new forms of contract.

The title also carries implications which might be questioned. Can contract forms influence the effectiveness of project management? Some argue strongly that the answer is no. They say it is the ability of people and the contractual culture - the way people and organisations behave - which are the dominant influences on management effectiveness. An alternative view is that both the form of contract and the contractual arrangement can influence both the contractual culture and management effectiveness. At the University of Birmingham work is just starting into the interaction of contracts and culture, led by a social scientist who has worked with the construction industry for many years. The research is stimulated by the New Engineering Contract but is not dependent on it. Interesting issues are already being raised. For example we recently debated the relative importance of two fundamental questions:

- how will the culture of contracts impact on the way NEC is used?
- how will NEC impact on the existing culture of contracts?

Will a new and radical contract form prove to be an instrument for change, as those of us involved in its development hope will be the case, or will only those parts of the new form be adopted which suit the prevailing culture? Early use of NEC suggests evidence of both of these trends. We need to know why and what are the influencing factors. There is a rich vein of research here.

Another implication of the title is that we know effective management when we see it. Some of the professionals would agree, yet in today's world subjective assessments rarely seem adequate. Targets are set and performance indicators are produced even in Universities. We are measured against student/staff ratios, publications per year per academic, research income per year per academic and space utilisation ratios to name a few. The question is whether effective management of projects can be measured. To do that we need to identify the relative weightings of the contributions which together lead to good management. If we can't do that then how do we know if any change - whether to a contract form or anything else - has improved management performance. I agree with Hugh Beale that precise measurement in a scientific sense is certainly difficult and I suggest it may be impossible. Some of the reasons for this are:

- variances between initial targets and final outcomes are always influenced by both unrealism in the target and management performance. These are extremely difficult to unravel, even if the project scope and objectives remains constant;
- we do not yet have measures to take account of the degree of difficulty of the management challenge although research has indicated the factors which influence it (e.g. riskiness, complexity, pace);
- people skills, of increasing importance in project management, cannot be quantified;
- projects differ in the relative importance of objectives such as Cost, Time and Quality and so the measurement of performance against any one target may be misleading;

Some ten years ago when I began to be convinced of the need for a new form of contract I would not have been able to produce the above list. As a result of R&D we understand the problem better but the solution remains intractable. It also shows that it is unlikely that we will have the ability in the foreseeable future to produce scientifically acceptable measures of project performance and we will need other means to assess the contribution which an innovation might make or has made.

Need for NEC

In order to make progress we need alternative ways of gathering and assessing data. Last year Hugh Beale indicated the way forward when he said that we need to "share existing knowledge and try to synthesise it." I would like to go back in time to see how that process worked, imperfect though it might have been, in reaching the conclusion that a new form of contract was needed.

One obvious place to start was to see if anyone had established a relationship between project management performance and contracts. I found that few organisations publish the results of project performance, perhaps not surprisingly given the many anecdotes about cost and time overruns. It is to their credit that the World Bank do so on an annual basis and Figure 1 is taken from the most recent of such reports in my possession.[3] Eight years ago their data was less sophisticated. Economic rate of return is an overall measure of project performance which takes into account the costs, timescales and expected benefit streams of projects. In simple terms the widening gap, shown on Figure 1, says that project outcomes have become increasingly difficult to predict and that more things are going awry, when compared with original expectations, than used to be the case. The management task is becoming more difficult. The view of the World Bank is that cost overruns and delays during construction have a relatively small influence on this widening gap but it is a big problem and any positive contribution to it is a welcome one. The sample size is large, with around 100 projects evaluated each year, and their synthesis of information should have some credence. Amongst the causes considered over twenty years they cite the following which are relevant to the structure of contract forms:

- scope changes
- inadequate design
- procurement problems
- unrealistic programmes
- inflation.

---

[3] Operations Evaluations Department, The World Bank. Evaluation Results for 1991. IBRD, Washington DC. USA. March 1993.

A more aggressive assertion of the importance of contracts to project management also came from the USA in the form of a Business Roundtable Report on the cost effectiveness of the construction industry. Their reports stated:

> *"Contracting practices are growing more complicated and few items have more impact on the cost of the project than the contractual arrangements. The returns can be impressive. A representative sample of major owners and contractors estimates that the way construction contracts are written can add about 5% to the cost of typical projects."*[4]

> *"Develop the terms of the contract to embody the interests of both owners and contractors, recognising the goals of each and the ability of each to control and reduce specified risks and costs. Develop a formal contracting plan in depth as a means of arriving at a logical approach to risk management based on the desired project objectives."*[5]

> *"In particular owners should avoid using superior bargaining power to enforce contract language that seriously conflicts with a contractor's goals. If they do, an adversary relationship may arise between the owner and contractor - a poisoned atmosphere in which the contractor may lose his incentive to try hard to meet the owner's objectives for the project."*[6]

---

[4]   The Business Roundtable. *More Construction For the Money. Summary Report of the Construction Industry Cost Effectiveness Project*. The Business Roundtable, New York, January 1983.

[5]   The Business Roundtable. *Contractual Arrangements. Report A-7 of the Construction Industry Cost Effectiveness Project*. The Business Roundtable, New York, October 1982.

[6]   The Business Roundtable. *More Construction For the Money.Op cit.*

During the late 1970's and early 1980's I was directly involved in two attempts to synthesise knowledge of contracts, both funded by CIRIA. The first, on Target and Cost-Reimbursable contracts,[7] revealed that no model contract form existed for target contracts (and none did until NEC was published) and only IChemE produced a cost-reimbursable form which was regarded with serious credibility by the industry. The second research study was on Management contracting[8] and a similar picture emerged, although the JCT Management Contract form was soon to follow. Both these pieces of research and a later study of an overseas contract for ODA[9] convinced me that contract forms can influence culture and behaviour. One observation which made a considerable impact on my thinking was that, where contractual arrangements and forms were designed to engender collaboration this was welcomed with enthusiasm by the more senior staff. I also noted that experienced, but less senior staff were not always comfortable with collaborative working and sometimes made little attempt to alter their normal behaviour patterns.

The body of knowledge on contracts had grown by the mid 1980's to the point where synthesis yielded the conclusion that change was needed. Industry had begun to question the uncoordinated way in which professional institutions produced contract forms. In particular there was dissatisfaction that single-discipline forms (aimed at civil, building, M&E etc), were not suited to modern multi-disciplinary projects. Differences in contract procedures and risk allocations produced the need for

---

[7]   Perry, J.G. and Thompson, P.A. *Target and Cost-Reimbursable Construction Contracts*. CIRIA Report 85. CIRIA 1982

[8]   Hayes, R.W., Perry, J.G. and Thompson, P.A. *Management Contracts*. CIRIA Report 100. CIRIA 1983

[9]   UMIST Project Management Group in Association with Martin Barnes Project Management. Songea-Makambako Road Tanzania. Phase II Target Contract Evaluation. Overseas Development Administration. August 1987.

different types of contract administration on the same project with the consequent inefficient use of staff resources and contractual problems at the interfaces of contracts.

There is not space to mention all the pieces of research and professional input which were synthesised into the need for a new form but I should mention two more. One was the almost overwhelming view that too much resource was being (and still it) consumed in dispute resolution, much of it at site level or just above. I note that Hugh Beale agreed that the contract form has a role in preventing disputes. He said:

> *"Contract law ... functions as a device for planning relationships to enable projects to be completed smoothly; contract (can be seen) as an enabling device, as a form of communication, as a systems of incentives. To me, the preventive role of contract is actually more important than its use in dispute resolution."*

It would be interesting to know what proportion of project managers share that view and whether the proportion of construction lawyers who also do is similar. Is it possible that the one figure could turn out to be the inverse of the other?

The final piece of knowledge gathering was a growing strength of reaction to many years of practical experience which found that programme clauses in traditional forms were toothless. The responses included requiring adherence to programmes submitted at tender stage with severe financial consequences if contractors departed from it. There are differing views about the benefits of this practice but my concern was that it might represent the tip of an iceberg, which was a proliferation of non-standard contract terms with varying degrees of robustness in both management and legal terms. There was (and may be still is) the danger of jungle warfare if the real needs of such clients were not met through model forms of contract, which carry the support of all sides of the industry.

The synthesis of these and other issues required judgements and, just as in politics and the social sciences, total unanimity was difficult to find and the case was not irresistable. Was the case for treatment so bad as to require radical reform or only bad enough to require fine tuning of existing forms? It was the newly formed Legal Affairs Committee of the ICE, of which both John Uff and I were members, which decided to develop a new form of contract. This was not the work of one person but I am sure John would join me in wishing to pay tribute to the central role of Dr. Martin Barnes, which was sustained over many years, in bringing the vision of a new contract form into reality. I use the word vision not in a 'Damascus Road' sense, but in the sense of forming a strategic view of a necessary change with clear objectives. This was a painstaking task involving many disciplines and required a huge amount of determination to bring it to fruition.

Structuring of Contracts

A number of questions which affect contract structure were considered in the development of NEC over a period of eight years. But the knowledge base was not sufficiently complete for the right answer to be clear, or to allow full confidence in the proposed solution. Four elements of structure are now taken for consideration, but there are many other areas for R&D.

a)   The Style of Language

NEC is written in relatively simple English with short sentences by comparison with traditional forms. We aimed to follow at least in spirit, the recommendations of the

Plain English Campaign.[10] Without searching, a supporting reference from an American Academic Lawyer[11] came into our possession

> *"We lawyers cannot write plain English. We use eight words to say what could be said in two. We use old, arcane phrases to express commonplace ideas. Seeking to be precise, we become redundant. Seeking to be cautious, we become verbose. Our sentences twist on, phrase within clause within clause, glazing the eyes and numbing the minds of our readers. The result is a writing style that has, according to one critic, four outstanding characteristics. It is: "(1) wordy, (2) unclear, (3) pompous, and (4) dull."*

One underlying hypothesis to NEC is that part of the cause of disputes lies in misunderstandings between the parties as to the meaning of the words in the contract. Sufficient people supported this for NEC to act on it. Nevertheless there are detailed questions for researchers to pursue.

• What proportion of disputes are either caused by or fuelled by lack of clarity in the lange of the contract?

• Is there evidence that certain styles of language help avoid misunderstandings?

• Will future contracts need to eliminate words or add words if misunderstandings are to be avoided?

---

[10] Cutts, M. and Maher. C. *The Plain English Story*. Plain English Campaign. 1986; *Small Print. The Language and Layout of Consumer Contracts*. National Consumer Council 1983.

[11] Wydick, R.C. *Plain English for Lawyers*. (1978) 66 California L.Rev. 727

It was Max Abrahamson who advised us that the English Language is a blunt instrument which is not capable of being used with the precision necessary to avoid misunderstandings in contracts. As you may know he has developed an even more radical form of contract (CCS) which is based on the principle of flow charts and is almost devoid of sentences.

- What legal issues arise with a contract form which looks like a flow chart?

b)   Forms for Minor Works

For years the accepted wisdom has been that minor works need minor forms. NEC is both shorter and simpler than traditional forms and the view of those who developed it is that it is suitable for work of any value. Nevertheless Eskom, a major South African client involved in electricity generation and transmission, has now placed over 700 contracts using NEC, a large proportion of which use Eskom's own Minor Works form, based on NEC. The use of NEC in Belize on a small road project using local contractors has shown that some small contractors lack the expertise to comply with all NEC procedures. For example they may not be able to produce programmes without assistance. There seems to be a need for further research and development on forms which suit the capabilities of small contractors, including those found in developing countries. Questions which intrigue me are:

- Can 'minor works' be defined?

- What factors govern the decisions to use minor works forms in practice?

- To what extent are existing forms for minor works effective in meeting the needs?

- Is the value of the work or the expertise of the Contractor more important in the choice of a minor works form?

- How different would a form for minor contractors look from a form for minor works?

c)   Adversarialism

Sir Michael Latham[12] has recently said, (in his Interim Report for Government and Industry on the UK Construction Industry):

"I have found a general mood for change. It is widely acknowledged that the industry has deeply ingrained adversarial attitudes. Many believe that they have intensified in recent years. There is also general agreement that the route of seeking advice and action from lawyers is embarked upon too readily. While a relatively small number of these legal disputes actually reach formal Court hearings, the culture of conflict seems to be imbedded, and the tendency towards litigiousness is growing. These disputes and conflicts have taken their toll on morale and team spirit. Defensive attitudes are common place. A new profession of "Claims Consultant" has arisen, whose duty is to advise some participants in the construction process how they should to make money out of the

---

[12]   Latham, Sir Michael. *Trust and Money*. Interim Report of the Joint Government/Industry Review of the Procurement and Contractual Arrangements in the UK Construction Industry. December 1993.

alleged mistakes or shortcomings or other participants. While clearly the existing culture of claims provides its own justification for such services, it is difficult to imagine a starker illustration of adversarial arrangements within the construction process itself."

This is a severe indictment of the industry and the professions who serve it. It cries out for radical reform. Has Sir Michael already accumulated a sufficient body of evidence through his extensive review process or is more needed, and is his synthesis?

- Can the assertions be supported by data?

- If the assertions are correct what remedies should the legal profession develop?

- If collaboration is the opposite of adversarialism and is to be encouraged then what problems are known about commercial arrangements based on collaboration?

- What balance need to be struck between collaboration and motivations for efficiency?

- Has NEC hit the right balance and, if not, how can it be improved?

d)   Dispute Resolution

This is an area of intense current interest with changes of practice occurring relatively quickly. Two examples are the increasing interest in and use of ADR and the elimination of Arbitration from contracts by some clients who view arbitration in practice as little different from litigation. It seems likely that future editions of NEC and indeed other contract forms will need to revisit this topic.

The debate will need to be informed by more research than appears to me to have been done in the past.

- Is there data on the relative costs and durations of arbitration and litigation?

- If arbitration is eliminated from the contract what are the consequences - both management and legal?

- Is there data on the relative merits and demerits of the different forms of ADR?

- It has been asserted that some types of disputes benefit from instant resolution whilst others are better left until the end of the contract. Are the reasons for this managerial or legal?

- Can different types of dispute be categorised in terms of the most appropriate dispute resolution procedure and, if so, how can this be incorporated into the structure of contract forms.

The topics I have considered can be added to and improvements can be made to the questions I have posed. But I believe they are all relevant to the structure of contracts in the future and they are all aimed at contributing to improved effectiveness of project management.

Concluding Remarks

Research in law often emerges as an individual view based, no doubt, on very careful reading of the cases. This is often difficult for the outsider to assess, especially if individual experts differ in their interpretations. The Centre of Construction Law and Management is ideally placed to develop new methodologies for research in construction law which might focus more on the gathering and analysis of data and opinion from a wide range of sources and from different professions. There is a difference of culture and methodology between law and the other disciplines involved in our subject such as Engineering, Management and the Social Sciences. If multi-disciplinary collaboration in research is to work more effectively, it is important to develop research methodologies which bridge the current gaps between the related disciplines. I hope that I have demonstrated the enormous potential for R&D in construction law and management, which can provide a service of great value to the construction industry and its clients. The Centre is exceptionally well placed to take a lead in these matters with the strong contacts it has generated through past students working in the industry, through the legal profession, and also through the many professionals who support the work of the Centre.

17. Risk, management, procurement & CCS

Max Abrahamson

Synopsis

This paper is a brief 'trailer' for the Construction Contract System (CCS) that is the result of contract drafting and analysis of risk management from 1970, participation in the conception of the NEC and on its ICE Working Group, and further work since.

Purpose

> 'If you have them by the [obscenity deleted] their hearts and minds will follow' (attributed to a resigned U.S.President). ... What I may call the testicular view of the construction industry is not working and cannot work, because the anatomy of construction is so varied that it is impossible to obtain a secure grasp .... if you live by the contract you are likely to die, or at least become extremely sick, by it'.[1]

The less than glowing health of construction has been confirmed authoritatively by the Latham Report[2], which describes failure to an extent that occurs only when a regulatory system is encouraging the

---

1 *Risk Management*, [1984] 1 ICLR 241 at 264
2 *Constructing The Team, Final Report of the Government / Industry Review of Procurement and Contractual Arrangements in the Construction Industry*, Sir Michael Latham, HMSO, January 1994.

tendencies it is meant to discourage. That is why the following tools
and methods are necessary.

The Construction Contract System ('CCS')

This system consists of an agreement, schedule, guidance notes
including an introductory tape[3]. The chapter and table below are
taken from the single CCS schedule, which in 14 chapters provides
lump sum, remeasurement, cost plus, design and construct main
contracts for building and engineering. The schedule will be
incorporated into subcontracts and consultancy contracts.
Supplements will fill the framework already included for
management-only contracts, and will add process engineering and
BOT. The very short CCS agreement emphasises that 'the parties
intend these contract documents to receive purposeful and
businesslike interpretation'. This is one of many cases where CCS
takes advantage of established law, which will apply that agreed
intention with the help of other CCS tools.

Schedule structure

The schedule is structured to carry the extra burden that the law
places on contract communications by making them binding, without
using language unfamiliar to those it is communicating with. The
CCS introductory tape gives examples of how the structure draws
attention during drafting to gaps and conflicts; allows simplicity
without leaving problems to disputes after contract; adds precision;
obstructs loophole engineering. The following are only a few of the
tools used:

---

3   Full copies are available from the author, at 100 New Bridge Street,
London EC4V 6JA, fax 071-919 1999.

Standard & options

The schedule can be used as printed except for completing a few marked items (for example the completion period at chapter **3** item 15, below). Or the schedule can be customised for special project features.

Checklist

The schedule is itself a checklist for attending to project features before contract, and for management communications and decisions after (mainly by the graphics, below).

Access

The schedule's headings and arrangement give the essential overview. It can be read through up to eight times in the time needed to read a traditional form, still remembering the beginning when the end is reached. A one page index locates details, highlighting the items where special meanings of words and phrases are defined or identified (the meanings adjust to different forms, such as 'complete' to 'completion'). Other tools below also make for easier access, learning and confident use, to reduce the errors in applying contracts, which cause a high proportion of disputes.

Actions & items

As shown, the schedule separates into numbered items the actions required or permitted from the parties. In a list made by indented items or by commas in the same item, each component applies to the subject matter of the list, for example items 16 to 23 apply to define completion in chapter 3, opposite, and both actual and possible defects apply under item 17. In that way, listing supplements punctuation and reduces the ambiguities of 'and, 'or'.

Itemising and listing also help to locate unavoidable qualifications and other complications in their right places, and allow precise cross-referencing by number to individual items in the schedule and, for example, in notifying delegation of the OR's powers.

Words

A wide word usually has a large area of 'fuzzy' meaning. If narrower words are added to limit it, each will have less fuzziness but there will have to be many of them. In construction contracts where technical and payment documents more than double the number of words in the 'legal' documents and many adversarial words may be added to each contract, ambiguities inevitably stack up like tolerances that never cancel out. Therefore, CCS in 5000 words says only once what is needed, without labouring the obvious or paraphrasing general law that applies anyway. Such brevity helps other CCS methods, and vice versa.

Graphics

The graphics (as legally effective as contract drawings) focus attention individually on:

- Project features, by grey optional items, rectangles and 'links'. Words inserted in a grey rectangle replace grey words on the left. References inserted in a grey 'link' show where another contract document completes or alters the item.
- Timing, by arrows as explained in a key (even the best advocate may find limited room to argue about an arrow).
- Management, by the wording on the right of the items.

Tables are also included, for management, insurance, etc.

Square brackets enclose 'reminders' of the purposeful and businesslike interpretation emphasised by the agreement. A mere reminder need not be repeated everywhere similar to avoid the legal presumption that a change of wording is a change of meaning. So because one item says that the Owners Representative may delegate '[all or any of his] actions' does not imply that under another item which says simply that the contractor may delegate his actions, they may only be subcontracted all together.

Parts

Modular 'Parts' add more flexibility to the schedule without many new words and complications. Parts are created by reviewing the schedule before contract, using the specification to define those required, and for each one adding to the schedule copies of its relevant pages with options, graphics and tables specially completed, with all incidentals attended to. Parts can be used for example to schedule different designers, design liabilities, completion and remedial periods for mechanical and electrical compared to civil elements of the Works; free issue of some materials by the owner; facilities from the contractor to public utilities and others; to define milestones with separate completion dates, payment, bonuses and agreed damages if desired (for which a table is included).

Software

CCS is designed to use the help at last available from electronic word processing against old problems of legal word processing. CCS software will add further simplicity and safety in choosing options, completing tables and coordinating Parts. It will pop up the special contract meaning of a word wherever it is used, and the listed extracts from specifications and drawings beside each grey link; extract special checklists of e.g warnings required by the schedule or specification. It will identify intended connections in the contracts documents and management information, and during drafting warn of risks of unintended connections.

Extensions

Specialists are free to use those methods in their technical, management and payment documents in and under the contract; extracts from the schedule will be published to supplement non-CCS contract forms, particularly on management (below); access to background law and general information will be organised around the schedule. The system is also adaptable to other legal systems, as it concentrates on principles and actions that have to be similar everywhere.

Owner's Representative

CCS does not reduce the status or effective functions of the owner's representative (OR) compared to the traditional contract Architect or Engineer. The OR is merely relieved from rights and duties that were safe only when contractors had few functions and fewer rights. The OR does not have to 'approve/accept' or reject contractor's planning in advance without full information or effective powers of correction, and risking claims against the owner and himself whichever he does. Instead, he intervenes only when something is or may be wrong, usually with a request and opportunity to satisfy it, not a peremptory demand, and with enforcement powers if needed. The contractor has the freedom and duty to proceed unless and until the OR intervenes.

Time saved in this way from the traditional built-in contract confrontation may be used by the OR (and his delegates) on site inspection and investigations. Contractor may use it on planning and getting the works right, not merely past the OR (and avoiding later liability). It is established that quality control, with personal satisfaction from achieving it, is more likely where there is no temptation to rely on or blame outsiders. The schedule does not include parties' promises to cooperate which it could not enforce, but cooperation is more likely when it cannot be rewarded with liability for the other's problems. Compulsory advance judgment by one 'partner' of each action by the other is hardly conducive to 'partnering'. The schedule uses 'warnings' particularly (below) to strike a balance in individual cases between ensuring that the Works benefit from all the available expertise and allowing each party to concentrate on his own specialty.

A member of the public suing the owner and OR for injury or damage actually caused by the contractor will no longer be able to rely on contractor's drawings marked 'approved' or 'accepted' by an owner's representative so as to imply that he sanctioned them as comprehensive and safe for use.

Timetable and Construction

The agreement makes the CCS guidance notes admissible in legal proceedings for background information but not as contract words supplying more ammunition for the disputes business.

3   TIMETABLE & CONSTRUCTION

1   Date

Safeguards

2   Party provides his original billed securities, copy insurances.

3   Owner makes any billed advance, in exchange for any repayment guarantee.

Contractor's 1st & regular reports

4   Contractor reports this information.

5   Contractor reports this information as of every   day between 1 and 22;

6   day between 22 and 32.

Site

7   Owner gives the contractor possession of this site,

8   this right of way in and out;

9   with this sharing.

Progress

10   Contractor supplies and constructs the Works

11   including:–

12   Complying with performance specifications.

13   Making suitable incidental choices where the contract is silent.

14   Providing fabrication, other working, drawings and specifications.

Due completion

15   Contractor completes within this period,

16   subject as here,

17   The Works,

18   with no defects, substantial possibility of defects, of which he knows,

19   unless his completion report shows that

20   their existence, investigation, remedying,

21   will not substantially reduce enjoyment, value, of the Works.

22   Tidying and vacating the site.

23   Supply of this user's maintenance, operating, information.

24   This completion reporting.

25   A 'defect' is Works:–

26   Outstanding at 13.

27   Containing contractor's breach.

Methods, programme.

5/6 Update of all contractor's information; account.

Instalments 4T.

Delay to completion.

The following are samples of the guidance notes on the extract from chapter 4 printed opposite.

1   'Date .... contract was [first] made'

This date is the date of an unqualified acceptance of the tender; if none, of the agreement. For certainty a specific date may be substituted in the grey rectangle, specially if a letter of intent was issued before the full contract was made.

2   Billed securities

See index for the guarantees billed.

The performance guarantee will end at the final account 3.32, except for claims already made, and may be reduced for Parts completed early. The referee has some control over calls made on it.

Owner cannot stop payment under letters of credit by dismissing the OR or preventing his approvals needed for the documentation; the contractor collects unless the OR intervenes positively, and the referee can prevent him going that far without reasons.

Mobilisation

Mobilisation that is clearcut for the particular Works, by specifying the contractor's resources, temporary works, required on site, may be defined as a Part with its own completion date (and payment, if desired).

7,8   Site and access details

For example, it may be specified that possession will be shared of some of the site as working space; whether access is public or private, who shares it and for what purposes.

9   'with any requirements, restrictions, here'

For example, requirements for site security, restrictions on working hours.

10 - 'Contractor .... constructs .... including'

Clarifications of 'construction' follow in 12-14 because in law the contractor guarantees that his construction will be suitable for its purpose even if he takes less responsibility for his Works design (**2**.13-15).

15   No completion period entered

If this arrow head is left empty, general law usually entitles the owner to any actual losses he can prove were due to the contractor failing to complete within the time that was reasonable in all the actual circumstances.

Contractor's control

Contractor is allowed to manage how he achieves completion in time, without indefinite requirements like 'regular progress'. But the owner can create milestone Parts with individual completion periods, and if desired individual payments, bonuses, agreed damages, and giving rights of termination for delay.

17-20   'defects .... not substantially reduce enjoyment, value'

Owner may use the link provided to supplement this definition - e.g specify which internal and external 'finishing' is required; limit 'tidying and vacating the site' (22) for completion of a Part; include the connection between two Parts in completion of one and exclude it from the other.

Payment

CCS has two payment bills. The priced bill is the standard for the original Works and some changes, with the costs bill for all other changes. But the costs bill includes an option to apply it to the whole Works and omit the priced bill altogether. The division between the bills minimises the room for distortions from mixtures of omissions and additions; rates, costs, percentage additions, fee. It encourages efficiency in performing changes, without perpetuating artificial pressures at tender. Particulars about targets, retention, fluctuations of prices and currency, VAT, etc. are entered (by the owner or tenderers) in the relevant bill so that they can be related to individual entries and groups of entries.

The priced bill provides pricing options for all the standard types of contract by a choice from defined units - the whole Works as a single unit, making a pure lump sum contract; units of volume, time, number, making a bill of quantities or full remeasurement form; activities for an activities bill. In the 'costs bill' the contractor will price listed cost units, and a lump sum or percentage fee which covers all costs not separately billed. The costs bill will include tables where the contractor gives particulars about his resources and his costs of using them. He must also complete copies of the tables for any additional resources needed for changes.

Management

> *"I am trying .... a management manual .... given contractual force"*[4].

Those efforts have proved in practice that more refined contract tools are needed to help management. Consequently, one schedule chapter provides a framework for management - meetings,

---

4   *Risk management, Op cit.* p.257.

reporting, warning, investigations and coordination - with the help of entries on the right of the individual items to show in context what must be communicated and (using colours, missing opposite) by whom.

That framework deals with management that most concerns both parties, particularly coordination of purposes (since it may be 'good management' for a contractor to create disputes from which he may profit). It does not presume to teach the contractor his own management, but requirements may be added by the owner (for which he will then have responsibility). The contractor should find it easier to use (and to tender for using) his own management methods within that framework, rather than unfamiliar methods dictated in the contract, and less easy to disown them if anything goes wrong.

Meetings

Accepting that a contract cannot compel constructive discussion, the schedule facilitates meetings of the right people, on agendas checked for completeness against the whole schedule, using scheduled information, with help available from the referee, and with little danger that responsibilities are transferred unintentionally by discussion. A mechanism is included for either party to call meetings and invite particular representatives of the other. That facilitates meetings by senior representatives of both parties in time to stop problems from escalating, and gives the contractor access to the owner more directly than through the OR.

Reporting

The information to be reported is scheduled in context item by item to avoid generalisations and ritual. The contractor mainly gives information that he needs anyway for his own purposes. The table opposite requires reports to be referenced to the schedule so as to take advantage of its organisation.

Takeover
After completion:—

28 Owner takes over the Works.
29 Contractor may refuse OR's change order unless about:—
30 Defects.
31 Minor variation to the Works.

After the completion period the owner may take over incomplete Works.

Remedial period[s]
After completion,
the contractor:—

34 Warns of defects for this period.
35
36 Promptly remedies each defect within 15, 28.
37 Repeats completion tests as relevant.
38 After 28-31.
39 Contractor finally reports.
40 and may not claim any sum not included in an account
41 except an indemnity under 10.
42 If the contractor causes a defect by 32.
43 30-33 apply for this period also, not this [35/36].
44

Date of completion;
successful completion test results;
each defect, substantial possibility of defect, with remedial investigations, methods, programmes; material, equipment, remaining on site for 37/38.
access requested.

[7] Correction to completion report.
Objection to access requested.
Refusal.

Date for takeover.

Each further defect.
Remedial investigations, methods, programme; access requested; successful repeat completion test results.

Objection to access requested.

To avoid requiring information before it can be firm, the owner may enter in the grey rectangles in that table periods of say, 'weeks' for the first '56' days, to cover mobilisation, and then 'months'. The contractor is given extra incentives to programme. For example, the bills include options by which the owner pays before completion only for contractor's progress when his programme has been adjusted for it. The contractor is not penalised, only losing cash flow while and so far as this information is missing. The information is useful for contractor's planning, and to compare planned and actual efficiency and progress.

Warnings

Early warnings are scheduled (in italics) for specific items. Their contents are clearcut and brief but enough to alert the recipient (after using the schedule as a checklist if necessary) to protect his interests and perform his responsibilities e.g. by calling meetings, investigating, ordering or proposing changes, mitigating his breach.

Co-ordination

Dangers are minimised from traditional divisions between design and construction and between information from each party in the contract and supplied later. Contractor is mainly responsible for coordinating his information with the owner's, because it is he who uses all that information in his construction.

Management options

Most types of management-only contracts are available by a few choices under one chapter. The first choice is whether contractually to connect direct to the owner a construction manager and each of several direct 'Works' contractors, or only a managing contractor who will then be connected to subcontractors. In either case all connections are made using forms derived from this schedule. The other choices are the Parts into which to divide the Works for convenient packages, possibly including Parts for construction, supply of equipment, etc, by the managing contractor. Excerpts from the schedule can be used for several levels of coordination, even to coordinate several specialties using different contract forms on the same project.

Tender management

The schedule reduces the work of both parties at tendering and the room for mistakes. After reviewing the schedule the owner will include in his invitation marked up copy pages or item numbers identifying the information he requires for pre-qualification or with tenders, and any sole or alternative choices left to tenderers, for example the completion period, insurance excesses. Pages or numbers can similarly be used to qualify a tender, and to help incorporate results of negotiations into the contract more precisely than by attaching minutes and correspondence.

Changes

To maximise the market for construction, changes on behalf of the owner must be allowed after contract, although like other industries construction would be easier if customers' demands did not change. But the management framework minimises the problems for both parties. All 'changed circumstances' that are to the owner's account are dealt with together in one chapter, which incorporates refinements to principles of risk allocation previously suggested[5] that have been shown to be needed by 20 years more testing in practice.

The essential justification for sharing risk remains that no one can know in advance where the effects of undue risk on the construction industry and individual members will strike, or whether it will cause

---

5 *Contractual Risks in Tunnelling*, (1973-74) Tunnels and Tunnelling, p.587 - 598.

minor or catastrophic reduction in quality of public and private works, personal injuries, environmental and other damage, or disputes. The more owners require contractors to gamble the more they are gambling themselves. The management framework reduces the room for abuse of the scheduled risk sharing, and for the fear of abuse that might otherwise cause an owner to delete some of his risks. The schedule also allows for contractor's savings.

Warning and reporting are used to promote agreement in advance on delay and cost due to an owner's change. Agreement is not limited to all or nothing: each party has an incentive to agree as much, as soon, as practicable, for more certainty in his cash flow and planning - 'minimum, maximum, final, whole, partial'. Contractor knows that the less precision and completeness in his information and the wider the gap between minimum and maximum the less chance of agreement. OR may at least agree a minimum increase in payment or time in return for the contractor agreeing a maximum.

Delay and Disruption

The schedule methods help each other against complex problems, such as determining what has not happened - the progress the contractor would have made and costs he would have incurred but for delay and disruption. So far as practicable that is deduced from facts, not hopes on paper. For that purpose the programming described in the table opposite fixes attention on the first 'use date' from when delay to completion and cost will begin if information is absent or altered later, fixed so far as practicable from the contractor's performance, on or off site, open to the OR's inspection. The removal of 'approvals' and 'acceptances' and the fact that the contractor has no reason to wait for OR's requests that, specially if he performs properly, he may never receive, reduces the room   for disputes about whether waiting for them or the contractor's own inefficiency was the effective cause of delay.

Disputes

By the chapter on disputes -
> '9   Party may invite the referee to:-
> 10   arrange, settle arrangements for, participate in,
> 11   a meeting, investigation'.

Referee may also require security for owner's payments; if they are late he may release existing retention early and stop future deduction, to help the contractor's cash flow until the owner's payments catch up.

For disputes, the referee is mainly a conciliator, not a decision maker, for the overriding advantage that the parties can explore solutions of all kinds with his help, without having to put every conceivable argument at least once and watch every word because he is going to make a binding decision. It should be easier to obtain people suitable for this job than for temporary decision making, and their help should be more acceptable to the other professionals on the Works. A binding adjudication by the referee is possible only in listed cases, which are clearcut to avoid disputes about whether or not they apply, and on which a final decision can be quick (without being crude) in order to maintain the 'life blood' of the parties. CCS procedure helps the referee/arbitrator to apply to each of the (usually many) questions included in a dispute the method of prevention and resolution best suited to it.

Enforceability

> *".... if any of us has a choice between responsibility and survival, of course it is survival that usually will win. That is why, for all its faults, a legal framework is necessary"*[6].

---

6   *Risk Management, Op cit.* p.264..

The above CCS methods are designed to do their management and other preventative work without sacrificing the enforceability in law that is the most user-friendly function of a contract because it reduces the benefits to either party from unfriendly use and therefore the need for actual enforcement.

Revolution ?

'[In 1936[7]] *Lord Atkin said this:*
'it is difficult to understand why businessmen persist in entering upon considerable obligations in old-fashioned forms of contract which do not adequately express the true transaction'.
Nearly sixty years on little if anything seems to have changed.' Lord Justice Saville.[8]

'*He that will not apply new remedies must expect new evils; for time is the greatest innovator.'* Francis Bacon.

Viewed in the light of the exceptional importance, complexity, dangers and permanence of construction, CCS will be seen merely to build on past progress of construction and law as necessary to do justice to both, and to be revolutionary only if revolutions already affecting them are overlooked[9].

---

[7]   *Trade Indemnity Company Limited v. Workington Harbour and DockBoard* [1937] AC 1, 17.

[8]   *Trafalgar House Construction (Regions) Ltd. v. General Surety and Guarantee Co. Ltd.* (1994) 66 BLR 42, 49

9   I am very grateful to my colleagues in the Institution of Civil Engineers' Working Group on the New Engineering Contract for help and encouragement with my efforts to continue further along the contractual path we have chosen.

18. Process Plant Contracts

Gordon Bateman

Synopsis

This paper outlines some of the principle features of process plant contracts, and how they differ from other construction or provide-and-erect contracts. The difficulties of guarantees of the chemical or biological process are noted. The inter-relationship of risk allocation and payment basis is reviewed. Some of the provisions of the Institution of Chemical Engineers (IChemE) forms[1] are outlined, including:

- the basic level of liability of the contractor and how that relates to the basis of payment;
- the approval of drawings; and
- the management of variations.

Process plant contracts in the water industry are reviewed, and in particular the experiences at Thames Water. Choice of a particular form of contract is only one part of the overall procurement strategy that an organisation might adopt. The paper concludes with some comments on the *Constructing the Team*.[2]

---

[1]   *Model forms of Conditions of Contract for Process Plants; Lump Sum Contracts, 'The Red Book', Revised 1981; Reimbursable Contracts, 'The Green Book', 1976, revised 1992;* Institution of Chemical Engineers, 165-171 Railway Terrace, Rugby, CV21 3HQ.
[2]   Latham, Sir Michael, *Constructing The Team, Final Report of the Government / Industry Review of Procurement and Contractual Arrangements in the Construction Industry,* HMSO, London, 1994.

Process Plant Contracts

The Works

It is important to recognise that a process plant is far more than a collection of civil, mechanical and electrical engineering components. A process plant is designed to facilitate a chemical or biological process. However good the individual elements of the plant, if the process does not work the plant is of no value to the client.

In general terms a process plant contract has three phases:

- design and construction of the plant

- commissioning and setting to work of the process

- optimisation of the process.

Each of the three phases requires a separate set of tests to demonstrate satisfactory conclusion. These and their purposes can be summarised as:

Completion tests[3] : has the plant been designed and constructed such that the individual elements work satisfactory and therefore the process is likely to work?

Takeover tests[4] : does the process work?

Performance tests[5] : how well does the process work?

---

[3] Green Book, cl. 32; Red Book, cl. 33
[4] Green Book, cl. 33; Red Book, cl. 34
[5] Green Book, cl. 34; Red Book, cl. 35

Completion Tests

Completion tests are in practice similar to the tests which would be undertaken under a civils or simple mechanical and electrical (M&E) plant contract. The tests are to establish that the civils elements have been constructed properly, for instance watertightness, and that the M&E plant operates, for example the plant rotates when switched on. Only if all the individual elements of the plant work, *i.e.* the completion tests are passed, are the takeover tests attempted

Takeover Tests

The takeover tests will follow a period of setting to work, which will include time to allow the treatment processes to become established. If the takeover tests are satisfied, the plant is taken over by the purchaser. At this point the process has been shown to work, and the purchaser can have beneficial use.

Performance Tests

Having taken over the plant, the purchaser carries out the performance tests under the supervision of the contractor. Performance tests are only included in the contract if the contractor is required to provide specific guarantees in respect of the plant's performance. Inevitably, specific guarantees lead to damages if such guarantees are not achieved, and the IChemE forms envisage liquidated damages for failure of performance guarantees.

Calculation of such liquidated damages can be difficult. Perhaps a simple example of the problem is where the contractor, at the time of making the contract, states the power consumption of the proposed plant. Liquidated damages of £x per kW, where £x is the net present value of a stream of additional power costs over the next several years, would be a solution. Later when the plant is in use the power consumption can be measured and compared to the guarantee.

Agreement as to exactly what the guarantee covers and measurement of the guaranteed parameters is often significantly more difficult. Even the above seemingly simple example has difficulties: is the plant being run at the output level envisaged in the contract? If not, the power consumption will be different anyway. Other characteristics are more difficult to assess, and in the water industry there are particular difficulties in relation to quality. For example the feedstock, raw river water or raw sewage, may have different characteristics from those envisaged in the contract. Although the purchaser is to carry out the performance tests, if a test is unsuccessful the purchaser is required to allow the contractor to make adjustments to the plant to improve performance. Damages become payable if the plant then fails to meet the guaranteed standard of performance within a defined period.

The IChemE forms require that use of the plant is secondary to the contractor's need to shut down the plant for adjustments to improve performance. Such priority may not always be possible in the water industry, where service to customers must be paramount. This and other matters are dealt with in the Water Services Association's Form P,[6] which provides a set of standard water industry amendments to the IChemE Red Book. Form P, which was negotiated with the relevant water industry contractors, can readily be adapted for use with the Green Book. Indeed it has been suggested to the IChemE as part of their consultation for the new edition of the Red Book that some of the provisions of Form P are relevant to all process plant contracts.

---

[6] *Form P. Standard Amendments to the Model Form of Conditions of Contract for Process Plants for Lump Sum Contracts 1981 Edition*, Water Services Association, London 1991.

Allocation of Risk and Basis of Payment

There is a natural commercial law relating to risk: the more risk a party carries the more he wishes to be paid; or the less he wishes to pay, if he is a purchaser. At one extreme is the 'one page' lump sum contract with no facility for extra payment irrespective of difficulties encountered. The contract price being sufficient to cover the risk the contractor is to bear.

At the other extreme is the cost reimbursable contract where the contractor carries no risk, he is simply paid all the costs he incurs. Indeed, and this is the crucial point, he has no means of pricing for any risk. In practice a cost reimbursable contract is not used in a pure form, as there is no means for including the contractor's profit. At the very least the contract needs to be 'cost plus': either cost plus lump sum fee or cost plus percentage fee. The fee normally covers Head Office overheads and profit. Neither form of cost plus has an incentive for the contractor to manage costs, indeed the cost plus percentage fee has an incentive to increase costs.

Fitness for purpose or reasonable care and skill

Under a price based contract the contractor can have either a strict fitness for purpose liability or a liability of reasonable care and skill. The Red Book incorporates the former. Clause 3.3 provides that

> *"...the Plant as completed by the Contractor shall be in every respect suitable for the purposes for which it is intended".*

In contrast, the ICE Design and Construct form states at clause 8(2)(a) that the Contractor's design liability is one of reasonable care and skill. The background to the two forms is perhaps rather different. The civils design and construct contractor would probably employ a consulting engineer to do the design work. Such a consultant is unable to accept liability greater than that of reasonable care and skill. The draftsmen of the ICE Design and Construct form recognised that and chose not to leave the contractor carrying a risk that he could not pass on and which he was ill-equipped to carry himself.

Process plant contractors, however, often develop processes in a research facility and can therefore test and refine the process before marketing it. Such processes are often the subject of patents etc. and details may be subject to restrictions for commercial reasons. That, together with the fact that feed stocks are generally of a consistent quality, leads to an strict liability for the product being acceptable to the contractor as well as desirable to the purchaser.

Under a cost reimbursable form the only level of liability that is possible is one of reasonable care and skill. This level of liability applies to all aspects of the work, there being no facility to price any greater risk. For those only familiar with price based contracts, this has unexpected consequences in relation to the cost of rectifying defects during the defects rectification period. The IChemE Green Book states[7] that the contractor is to be paid the cost of rectifying defects unless:

(a) the defects are in work carried out by a subcontractor who is not eligible for additional payment for rectifying defects, *i.e.* the subcontract is a price based contract; or

(b) the contractor is negligent; or

(c) other special arrangements are stated in the contract.

---

[7] Clause 35.2

Payment Mechanisms

Payment mechanisms are very different between the two IChemE forms. For the Red Book a simple milestone or time-based payment mechanism, based on a proportion of the contract price. These mechanisms need to be worked out by the purchaser as the details of the payment arrangements are not set out in the forms themselves. Generally the IChemE forms leave much of the contract detail to the schedules which form an integral part of these contracts: for example, the Red Book has eight schedules covering both technical and commercial matters. The schedules cover matters which would be included in the Conditions of Contract, the Specification and the Appendix to the Form of Tender under an ICE6 contract.

The Green Book calls for a cost-checking system. It takes account of the time involved in such checking in the following way. Any monthly payment[8] consists of:

(a) the agreed estimate of the costs the contractor will incur in the month following the payment application;

plus

(b) the actual costs incurred in the month prior to the month of the payment application;

less

(c) the previously agreed estimate of the costs at (b).

Note that the month of the payment application is not mentioned.

---

[8]   Clause 39.2

Approval of Drawings

The Red Book contract Schedule 2, entitled 'Drawings for Approval', is used to set out those drawings which are to be submitted to the Engineer for his approval. There is the usual provision in the Conditions of Contract that such approval shall not relieve the contractor of his liabilities under the contract. Nevertheless, it is important that the list of drawings to be submitted is kept fairly short: it is a great temptation for the purchaser or his Engineer to want to check everything, in which case one has to ask why choose a design and construct form.

The grounds upon which the Engineer may refuse to approve drawings are set out clearly in the Red Book.[9] They are limited to occasions where a drawing does not comply with an express provision of the contract, or that it is contrary to 'good engineering practice'. Any "preference" engineering that the Engineer may wish to undertake is to come under the heading of Variations, unless it can be argued that the Contractor's design does not accord with 'good engineering practice'.

The provisions in the Green Book are very different.[10] Here the Engineer can disapprove a drawing and the contractor can dispute that disapproval only if the contractor can show that the Engineer's requirement is not in accordance with some express provision of the contract, or is contrary to 'good engineering practice'. The difference in approval mechanism reflects the different payment structure: a contractor on a lump sum must suffer the minimum of interference, while one having his costs reimbursed will have a more relaxed attitude.

---

[9]   Clause 20.6
[10]   Clause 19.3

Variations

Like any design and construct contract, the IChemE forms have a rather different approach to variations as compared with a construct-only contract. Under the IChemE forms a variation is a change to the specified requirements set out in the contract.[11] It is not any and every change that may be made, and certainly does not include every change that the contractor may choose to make to his design. Particularly because the Engineer does not know the full detail of the contractor's design it is not desirable for the Engineer simply to order a variation. Far better that he ask the contractor for his proposals for implementing a proposed variation. In the case of a Red Book contract, those proposals should be in the form of a quotation such that the cost to the purchaser of the variation is clear in advance. If that is acceptable, the Engineer can order the variation. The need to agree costs is arguably less relevant on a Green Book contract, although the purchaser will wish to know what costs he can expect to incur, both in relation to the disruption and to the work itself.

It may be that the contractor considers that he has identified a change to the original contract specification that will be beneficial. Such a change is a contractor's proposal for a variation, and if acceptable to the Engineer it can be implemented as a variation.

Limitation on the Contractor's Liability

ICE 5th Edition had no overall stated limitation on the contractor's liability, and no limit on the aggregate value of liquidated damages for delay. That is rather different from some of the M&E forms of contract. The difference reflects the risk.

---

[11] Green Book, cl. 17.1; Red Book, cl. 16.1

Generally a contractor prefers to be liable for liquidated damages for delay rather than to be liable for damages at large. In that way he knows rather better what are his potential liabilities. Similarly the more risk associated with particular work, the more the contractor wishes to see a limit on his potential liability, including total potential liability for liquidated damages.

A civil engineering contractor who can obtain his equipment, materials and labour from numerous sources will have limited concerns of his potential liability for liquidated damages is unlimited. If the contract includes M&E plant then he has greater concerns. The M&E plant will be on a delivery of many weeks or months. If the supplier fails to perform adequately the contractor cannot simply source the supply from elsewhere and not incur delay. If the contract includes advanced electronics, instrumentation and computer hardware or software then even more are the risks that if a supplier fails the whole contract may be very late. Hence the contractor's desire to limit his potential liability.

The IChemE forms provide overall limits both for the contractor's liability, other than issues for which insurance cover is to be provided, and liquidated damages for delay. Typical limits may be an overall limit equal to the contract price, and a total limit on liquidated damages (both for delay and failure of performance guarantees) of 20% of the contract price.

Process Plant Contracts in Thames Water

Historically the water industry was very reluctant to recognise its many sewage and water treatment works as process plants, particularly when considering contracts for their construction. This was certainly the case in the former Thames Water Authority (TWA). This lack of recognition was encouraged in the TWA, in that the various functions such as; civil engineering design, M&E design, procurement, construction supervision were kept in clearly defined 'boxes', and the co-ordination between the 'boxes' was at a very high level and hence very general and not project specific.

Normally civils works were fully designed either in-house or by a consultant, leaving the contractor to undertake construction work only. The approach to procuring capital projects was to split the works into numerous small- to medium-sized single-function contract packages. M&E work would be in many packages, often with the manufacturer of an item of plant responsible for the supply and installation of only his own manufacture. Hence plant selection was closely controlled by the client.

Many different forms of contract were used, with capital cost being the most controlled of the time-cost-quality trio. Quality was largely controlled:

(a) by the design being managed separately from construction; and

(b) by selection of suitable plant suppliers, facilitated by a separate contract for each supply.

This approach was general throughout the pre-privatised water industry.

Overall therefore the client carried the often complex issues of risk management and construction planning associated with the numerous contracts and their different provisions. Alternatively that risk management was undertaken by the client's consulting engineer. Frequently such risk management was not recognised by the client for what it was.

The Impact of Privatisation

Many changes in procurement practice took place before and after privatisation in 1989. Those changes which occurred before privatisation were either consequent upon the impending restructuring or perhaps were driven by the same factors, in part at least, as those which encouraged the government to privatise the industry.

Time was rapidly running out in the case of inadequate sewage treatment works, with legislation relating to effluent quality, which set latest dates for compliance. Capital expenditure programmes rapidly expanded. In 1988/89, the last full year prior to privatisation, Thames Water's investment programme was £191m. This rose to £240m in 1989/90 and to £418m in 1990/91. The level of investment has remained sensibly constant since then, and the recently announced 'K' value will allow for an average of around £420m pa from 1995/6 to 1999/2000.

It was recognised that the old management structure would not be suitable to deal with such a sharp increase in investment, nor was time or the resources available to continue with the old procurement strategies. Much of the detailed interface management had to be passed to the contractor. Further, privatisation had brought with it the freedom and responsibility to make long term financial and investment decisions.

At Thames Water the decision was made to centralise the Engineering Division at one location, subject to a few minor exceptions. Engineering was given a total project management focus with a project manager appointed to be responsible for each project. Dedicated resources covering all the functions necessary for project delivery, such as design, procurement, construction and project control, are provided to the project manager. In addition, two functional specialist groups in the Division have overall responsibility for providing resources to the project groups and for establishing standards and policies for all functional activities. Hence a matrix management structure was achieved which focused on project delivery.

The formation of the multi-disciplinary teams was just one factor which facilitated the recognition of the water industry as a process industry. While the overall investment programme was increasing dramatically, commensurate increases in internal resources were sadly not forthcoming. The lack of timeliness of projects designed by consultants, for reasons not necessarily the fault of the consultants, led to the implementation of design and construct contracts to meet the demand

First attempts at Process Plant Contracts

An early attempt in Thames Water was based on the ICE's early draft of their Design and Construct form, then a derivative of the 5th Edition. This lump sum form was used with limited success on two projects for the uprating of small sewage treatment works. Problems arose when it was identified that it would be beneficial if the contractors were to take responsibility for the quality of the treatment process. The works were recognised as a process plant, but the contract form did not address process plant issues.

A later example was the contract for the reconstruction of Camberley Sewage Treatment Works. Time to complete the work was very short, as the effluent quality had to be improved by a certain date, driven by legislation relating to discharges to rivers. Design and Construct was the only viable option, and yet the works had not even been 'scoped'. A cost reimbursable form was seen as allowing the necessary flexibility. That the works would be a process plant led finally to the use of the Green Book form. The two-phase approach, described below, was used, and the contract was considered most successful.

Changes in Procurement Policies

This early success led to greater use of the Green Book, including non-process applications such as the construction of the later stages of the Thames Water Ring Main using free issue tunnel boring machines (TBMs) and tunnel segments. Thames Water purchased three TBMs and contracted for the relocation from the Isle of Grain to Southall of an ex-Channel Tunnel segment casting facility.

What then was the thinking behind the more general choice of the Green Book? In considering the then current forms and the early successes with a new form, it was recognised that the older forms and strategies tended to lead to confrontation and adversarialism. In 1989 contingent liabilities on existing contract claims, on all forms of contract, exceeded £30m with numerous formal disputes and a number of live arbitrations. It was considered that so far as possible the contract should be structured such that the contractor's goals under the contract were the same as those of the client. In this way cooperation, not confrontation, would be encouraged.

In shaping the new organisation and its procurement policies a number of key decisions were taken. It was considered that Thames Water as an expert client should:

- always undertake its own project management;
- always manage its own construction sites;
- always do its own process design;
- always develop project designs to the Definitive Design Report or "30% Design" stage;
- establish design, procurement, construction and project control as primary functions in the client organisation, focused on a project management approach;

- generally not engage consultants in their traditional role for project work, as distinct from planning studies and the like; and
- let projects which were not designed in house as single packages with a single client/contractor interface.

Projects which (particularly the process and civils aspects) are fully designed in-house are also generally let as single packages, covering all construction and provision of M&E Plant, and thereby adopting the philosophy of a single client/contract interface.

Use of the IChemE Forms in Thames Water

Thames Water makes much use of the IChemE forms. In 1993/4 the total value of contracts awarded on the IChemE forms was £157m, including £125m for the two East London sludge incinerators. Further the Extended Arm agreement with Taylor Woodrow was based on the Green Book. The IChemE forms are used for a variety of process plants, including:

- sewage treatment plants;
- water treatment plants, and their associated waste water plants;
- ozone plant for advanced water treatment - both ozone generation and use;
- oxygen generating plant - to feed the ozone generating plant;
- granular activated carbon (GAC) regeneration plant;
- sewage sludge incinerators.

The Red Book is amended with regard to the strict liability requirement. As discussed above, Thames Water has adopted a policy of undertaking its own process design. With the exception of proprietary plant, for which a strict liability is required, the contractor's design of the plant is to be undertaken with reasonable care and skill, for similar reasoning as applies to the ICE Design and Construct form.

The Green Book is always used with a target cost mechanism, thereby ensuring that the contractor has a clear incentive to manage costs. While there is no real contractual need to separately identify the costs of a variation on a straight cost reimbursable contract, on a target cost contract the contractor is keen to have the target increased by the cost of any additional work. In a design and construct contract not all additional work is the result of a variation. It may simply be a change to the contractor's design: hence the potential for conflict.

The target cost approach is used by Thames Water:

(a) *where the plant is clearly scoped in the tender enquiry document*

In this case the tenderers are required to tender the target cost, along with certain fees, design, management, etc. The contractor's actual costs are later compared with this tendered target cost in assessing savings or overrun, and the sharing thereof between the parties.

(b) *where the plant is not clearly scoped and the contractor's design work will include much definition of the scope*

In this case the tenderers submit tenders for the estimated target cost, along with fees. The successful contractor may well not be the contractor with the lowest estimated target cost. At the end of the design phase, and before construction can commence, the Agreed Target Cost will be established by the parties. It is against the Agreed Target Cost that actual cost will be judged at the end of the contract, subject to an influence of the estimated target cost. This is to discourage very low estimates at time of tender.

Both IChemE forms are used in such a way as to encourage the contractor to have similar goals to the purchaser. Any client is concerned about time, cost and quality. These factors can be influenced under the contract as follows:

Time	- liquidated damages for delay
	- bonus for timely completion
Cost	- competitive tendering, including subcontracts in Green Book
	- target cost formula (where applicable)
Quality	- standard specification for workmanship
	- liquidated damages for failure of guarantees of performance

The long term nature of the relationship between the client and his contractors can be of major significance in achieving good performance. Such commercial factors are of more effect than anything in the contract, and were a major factor in the decision to move to a reasonable care and skill basis for the design.

Choice as between lump sum or target cost is largely dictated by the likelihood of change. Change is not easily managed under a lump sum form and a culture of 'no variations' has to be positively encouraged. However if changes are likely to arise:

- due to unforeseen conditions;
- due to the existing plant, which could not be inspected prior to contract award, not being as expected; or
- due to complex programme constraints, particularly where interfacing with the operation of an existing treatment works;

that is good enough reason to use the target cost version.

Any significant lack of definition of the plant is also a sound reason for using the two-phase target cost version. For the one-phase target cost version the plant should be scoped almost to the same degree as for the lump sum version.

The Latham Report

Constructing the Team recognises[12] that the process plant industry was not properly consulted. In view of the width of the building/construction industry, some limits had to be set, and one accepts that. To a large extent it probably does not matter that the process plant industry was not fully involved: the essential messages are there for all to accept and adopt.

Many of the recommendations of the Report have already been adopted by Thames Water, but there is no room for complacency and there is much to do to improve matters further. Reference has already been made to the recognition of the need for more cooperation and less confrontation. Forms of contract can go some way to facilitate that change in style and emphasis, but ultimately it is down to the people working on the contract and the approach they take. This is something that cannot be legislated for, it needs a 'hearts and minds' education.

Early use of the IChemE forms was most encouraging: no real arguments or problems around differences of view. Since then there has been a tendency to erode that happy position. Indeed, one or two contractors who were not involved in the earliest use of these forms have come to the same form of contract with a much harder, adversarial attitude. Those contracts have absorbed significant amounts of Thames Water's management's time. We

---

[12]   *Op cit.*, Recommendation 5, § 3.18.

need to understand why that different attitude was there. Was it simply the contractor's culture, in which case there is inevitably a reluctance to give them further bid opportunities; or was it that problems and costs out-turned in a different way to that expected, or that they had their 'B' team on the job and not their 'A' team, in which case we both need to understand the other's viewpoint and then give them a further opportunity.

Too much weight must not therefore be put on the particular form of contract other than:

a) where it dictates or reflects the basis organisation of the industry which is considered inappropriate; and

b) in recognising that a new form can help in signalling to all involved that a change of approach is required.

It is not necessarily the case that the New Engineering Contract (NEC) is more appropriate than the IChemE forms for process plant contracts. Whilst it is hoped that the second edition of the NEC will be closer to the Sir Michael's recommendations, it should be noted that the IChemE forms are already in line with the major issues in Latham, such as recognising the need to discuss and agree variations before they are instructed. IChemE forms also have the advantage of accommodating directly the three-phase structure of a process plant contract as set out above.

The language of the NEC of course is very different than that of IChemE forms, but a cynical voice within says: "what if one party does not do what the contract says he is doing." How does that compare with that party not doing what the contracts says he shall do? The NEC has been drafted essentially to cover the construction of a structure. Like the lost traveller who was told "I wouldn't start from here", it would seem that the IChemE forms are a better starting point for process contracts than the NEC.

This is not to say that Thames Water will not use the NEC in the future. But, having made a major investment in standard forms, the associated training, both of Thames Water personnel and of contractors, and the investment in terms of time and resources, which in part pre-dates the publication of the NEC, it will be some time before a move is made to introduce the NEC. What is important is that the essential messages of *Constructing the Team* are understood and applied when drafting the detailed mechanisms which are to be contained in a particular contract.

Ultimately contracts are dependent for their success on the people involved in drafting the contracts, and in undertaking them. Thames Water's approach is to work with its contractors in a positive way, and to continually consider how better this might be achieved.

19. Construction subcontracts

Will Hughes, Colin Gray & John Murdoch

Synopsis

This paper is based upon a study of specialist and trade contracting in the construction industry commissioned by CIRIA and undertaken by the University of Reading with Sir Alexander Gibb & Partners Ltd. The study was to provide guidance for the management of projects where much of the work is executed and designed by specialist and trade contractors (STCs). This paper describes a preliminary investigation into the nature and origins of specialist contracting, with a survey of the problems confronting STCs.

Introduction

All construction work involves either subcontracting or separate trades contracting. Therefore, development of the industry depends upon specialist and trade contractors. Herein lies a fundamental dilemma: the problems experienced by Specialist and Trade Contractors (STCs) are dominated by short term issues of business survival, whereas the development of an effective and efficient STC industry is a long term issue. Business issues are especially difficult for subcontractors because they are inevitably subservient to the financial, contractual and procedural systems imposed upon them by main contractors. In addition, main contractors may attempt to protect their own interests by transferring risk to their subcontractors.

An example will illustrate this. Specialist subcontractors are particularly vulnerable when a significant design contribution is required from them. Where such a subcontractor is appointed too late for its design to be integrated into the normal design development, the work can become the subject of contention because it is very likely to be out of step with both design and construction. Such contention is the direct result of poor project management, *i.e.* the lateness of the appointment. The consequential effects on the main contractor's programme can have a "domino effect" on the work of other subcontractors due to variations and subsequent re-programming. The main contractor then has little choice but to fight off claims by the other subcontractors, using any commercial or contractual means available.

Such a situation is a direct result of a general failure to understand and manage the complexity of the specialists' contribution. Experiences like this lead employers and contractors to protect themselves by including their own amended clauses in the contracts they sign. These are often interpreted as adversarial or onerous; typically covering issues related to payment, retention, liquidated damages, programme, set-off and attendance. Tackling these short term issues is mere fire fighting, and does nothing for the long term issue of developing a sound and competitive construction industry.

This example also illustrates the two categories of problem which always need separating. First, subcontracting is a mechanism for employing others without making them a part of one's own organisation. This is not a problem, as such. It is a common feature of modern commerce and need not cause any difficulty. It becomes problematic in the construction industry because it is so frequently combined with other problems. Second, specialist and trade contractors have particular skills and ways of working which make their role and contribution not only critical to the success of the project, but also very complex and difficult to manage. The complexity associated with the integration of STC work is not diminished by any particular contract structure. Their contribution is important, regardless of the type of contract. The critical questions to focus upon are about who actually undertakes work on sites and about how best to harness their skills so that a positive contribution is made to the construction process.

The rise of specialist contracting

Significant factors combined to accelerate the change from direct employment to subcontracting and to specialist contracting. These powerful forces apply in a wider context to all industries but have been particularly significant in shaping the trading patterns in the construction industry. The purpose of this section is to consider the wider, general developments and provide a context for discussing the specific developments in the construction industry. Two things become clear from considering these factors; first, why these trading patterns have emerged and second, why it is not plausible simply to avoid subcontracting and specialist contracting.

The causes of increased levels of subcontracting

The general increase in subcontracting has been driven by technological, political, social and economic change:

- As technology grows more complex, more diverse skills are needed[1]. One way of securing competitive advantage is by being the best in a particular technology. The changes caused by modernisation are

---

1 Lawrence, P C and Lorsch, J W (1967) Organisation and environment: managing differentiation and integration. Harvard University Press; Massachusetts.

irreversible. The construction industry has yet to accommodate them fully.

- The UK government since 1979 has been committed to encouraging an enterprise-based culture centred on individual initiative and reducing the influence of the public sector[2]. This encourages those who wish to start new businesses.

- Changes in patterns of work and career structures have led to expectations for more autonomy and personal control. Coupled with a favourable tax regime at a personal level, this has helped to drive more people towards self-employment and specialisation[3].

- In responding to fluctuations in the economy, particularly the recent recessions, firms have concentrated on their core businesses. Subcontracting enables them to respond quickly to changes in demand and gives firms much flexibility in changing the size of their businesses. Similarly, as interest rates have been the government's primary tool of fiscal management, firms have developed tight financial controls as a response to the rapid changes in the cost of borrowing[4].

Construction projects and capital spending have always been hit first as key regulators of economic activity. This has required the construction industry to adapt more quickly than most. The general trends noted above were given an extra push in the construction industry because of two events in the 1970s:

---

2 Conservative Party Manifestos, 1979-1992.

3 Handy, C (1989) The age of unreason. Arrow; London.

[4] Gray, C and Flanagan, R (1989) *The changing role of specialist and trade contractors.* CIOB; Ascot.

- The building strike of 1972 led to a disinclination for general contractors to employ labour directly, and a consequent growth in subletting of labour. This enabled contractors to reduce their dependency on trade union labour.

- In 1966, the government introduced Selective Employment Tax (SET),[5] which was designed to tax firms on their payroll. Firms wantd to minimise their tax liability, so the immediate consequence of SET was that contractors sought alternatives to directly employed labour.

- Althoug SET was repealed in 1972,[6] its effects upon subcontracting were compounded because of the building strike in 1972 which led to a disinclination for general contractors to employ labour directly, and a consequent growth in subletting of labour. This enabled contractors to reduce their dependency on trade union labour. Coupled with the effects of SET, subcontracting enjoyed a surge in popularity.

These two events were mainly responsible for a sudden surge of labour-only subcontracting, alongside the trend towards increasing specialisation. There have been additional influences at the project level, particularly the approach to design in the UK, but also the need to spread both economic and legal risk.

The emergence of nominated subcontracting

A feature of subcontracting peculiar to construction is the practice of "nomination" which has evolved to cope with four major issues;

- the increasing sophistication of construction,

---

[5] Finance Act 1966, s. 44; Selective Emplyments Payments Act 1966.
[6] Finance Act 1972, s. 122.

- the need to modify the main contractors' control over specialists,
- the role of the design team,
- the needs of clients.

These issues are expanded below. Nomination was implemented by representative bodies at UK industry level as a technique which sought to balance the competing demands in the construction matrix. The following discussion traces the context and the rationale behind nomination, explaining how each sector of the industry benefited, before turning to a discussion of the reasons for its current unpopularity.

a) Increasing sophistication

Increasing technological complexity has been matched by increasing numbers of specialist contributions to the construction process. This is because the use of new technologies needs highly developed skills and expertise. The process of technological development is often driven further by the increasing specialisation of those who deal with it: they become better at innovating. Even those clients and designers who do not innovate for the sake of it find that economy of the process and market demand prompt them to use these new technologies. This relentless progress modifies the very nature of the construction process. The traditional assembly process is changed by the need to start procuring specialist elements at an early stage.

Procuring these specialist components can take longer than the whole building. The answer is to specify, in advance, the installer of the components, leaving the main contractor no choice in the matter. The installer can then prepare the components so that they are ready for installation at the appropriate time. Failure to do this will result in the building site being kept waiting while the components are procured. The primary motivation behind nomination was the need to harness the skills of specialists before the main contractor was appointed. It is one of the strongest arguments for nomination and has helped to spur the growth of specialist subcontracting.

Specialists favoured nomination because it protected them from unbridled market forces by enabling them to compete on some basis other than cost. It also enabled them to develop strong and stable business relationships with regular clients of the industry and with certain consultants. In the event of main contractor insolvency, some construction contracts make provisions for direct payments to nominated subcontractors. Further contractual protection arises from the provisions for adjudication of disputes between contractor and subcontractor.

b) Main contractors' risk

Main contractors tended to favour nomination since the work of nominated specialists is not part of the main contractors' lump sum, but a cost-reimbursement element. In projects with a large amount of nominated work, the contractor is more a co-ordinator than a fabricator of construction work; and a conduit for specialists' money. There are contractual advantages for main contractors, especially under JCT 80. Examples include the fact that delays on the part of nominated subcontractors qualify for an extension of the contract period. (The same is not true of civil engineering or government contracts.) Under standard-form contracts, main contractors usually enjoy a cash discount for prompt payments and often deduct the discount even when payment is not prompt. The lack of involvement in the selection of nominated subcontractors is seen as a disadvantage by main contractors.

c) The role of the design team

Nomination was favoured by design teams because it enabled them to do two things. First, they could influence the quality of detail by using only their preferred specialists. Second, they could avoid having to detail or re-design work done by technically incompetent domestic subcontractors who had been selected by the main contractor on price alone. Such re-design work would rarely enable the client's design team to claim payment.

Cost consultants developed techniques of budgeting which encouraged competition among nominated specialists even though the main contractor was reimbursed for whatever was paid to such specialist subcontractors (prime cost sums). This allowed for continuing design development with the subcontractors, avoiding the traditional contractual chain. Also, since design liability did not apply to the main contractor, separate design agreements emerged to tie the specialist directly to the client. The significance of these features was a growing professional infrastructure around the nomination process, which would indicate a vested interest in retaining the status quo.

d) The needs of clients

Nomination protected clients' interests by providing some certainty of performance and by providing a direct contractual link (via the employer/specialist design agreement) with those who designed the specialist work. Many experienced clients have derived great benefits from developing stable business relationships with certain specialist subcontractors. Nomination enabled these clients to control the main contractor's selection process. Main contractors working for a lump sum stand to gain the most by competitively sub-letting their domestic subcontracts. Clients who wish subcontractors to be selected on criteria other than price can take control of the selection of subcontractors through the nomination procedure. This avoids the risk of choosing under-capitalised or inexperienced subcontractors.

The decline of nomination

The whole industry co-operated in the development and growth of nomination; it was popular. More recently, the practice of nomination has declined in popularity for several reasons:

- The complex matrix of contracts in nomination has been at the centre of some extremely difficult litigation. This has highlighted tremendous problems in the three-way relationship between client, contractor and subcontractor.
- The legal problems are especially difficult when a nominated subcontractor fails to perform properly or becomes insolvent.
- When the short-list of preferred specialists becomes very small, the lack of economic competition between them can unsettle consultants and clients.
- Worse, since the main contractor is reimbursed for all nominated subcontract work, the contractor's motivation to control expenditure might not be as strong as it would be for the priced work.
- Main contractors can face enormous practical difficulties because of the special nature of the relationships that specialists may have with clients and contract administrators.
- Main contractors have come to wish for as much control over specialist subcontractors as they have over their domestic subcontractors.
- Continuing and increasing exposure to liability of professional consultants has driven up professional indemnity premiums and there has been an increasing reluctance by consultants to accept liability for various aspects of construction work. By declining to nominate specialists, they avoid some of the liability associated with their design, co-ordination and performance.

For all of these reasons, the established patterns of nomination procedures are not being used. This leaves building clients vulnerable to the risks associated with lowest price bidding. It creates an atmosphere of tension and defensiveness that does not encourage best practice. The result is that the structure of the contractual relationships set up by nomination are no longer appropriate.

Alternatives to nomination

Disenchantment with the nomination processes has lead to a search for alternatives (such as "naming"). The alternatives may be watered down versions of nomination and as such they merely make the situation less well-defined, rather than offering clear solutions to the problems. Management contracting appeared for a while to be a better approach, but the way that it is used in practice is often as another version of general contracting. There are procurement methods that offer more effective techniques of integration, such as design-build and construction management. The former places clear responsibility for everything with the main contractor; the latter reduces the main contractor's role to that of an advisor and co-ordinator. These offer clearer contractual structures, but in practice the relative economic power of the client and design team is undiminished; procurement methods do not alter the relative sizes of the firms involved. Since even small quite modest projects involve subcontracting, the UK construction industry is now largely dependent on small specialist firms who do not have the economic strength to insist on terms of business that enable them to perform properly. Therefore, despite continuing developments in contractual and procurement techniques, there is still conflict between the engines of activity, the specialist contractors and their clients. The root of this is the fact that others are usually involved in the transaction between specialist and client.

Strategic issues for the industry

A healthy STC sector is essential to the continuing success and development of the construction industry. In some foreign industries the STC has been the focal point for improving efficiency and technical knowledge[7]. By contrast, the development of STCs in the UK has been conditioned by the factors outlined above. The constraints on the development of STCs have resulted in compromises that produce a far from ideal situation. The economic recession has reduced the workload of contractors and specialists. Competition is more intense now than any time this century. Specialist firms are being invited to compete for work on increasingly onerous terms as clients and contractors seek to minimise their own exposure to risk. The existence of myriad small specialist firms[8] willing to work under onerous circumstances has enabled and encouraged these practices. Some successful STCs grew into large firms, and some large firms specialised to become STCs. While not all STCs are small firms, it is the preponderance of small firms (with which large firms must compete for work) which largely dictates the patterns of business relationships in construction. Therefore, although the large STC firms have, in theory, the economic muscle to resist onerous contracts, in practice any attempt to do so will result in the work being awarded to smaller, more acquiescent firms.

Large firms often employ small STCs as a defensive strategy. This strategy has little to do with the reasons that the small firms exist in the first place. This leads to a conflict of motivations: some STCs trade in this way so that they can better develop the technical competence that provides their competitive edge; others trade like this for purely economic reasons. Those who

---

[7] Gray, C and Flanagan, R (1989) *op cit*

[8] Government Statistical Service (1991) *DoE housing and construction statistics.* HMSO; London.

employ STCs do so either because this is the best way to harness technically sophisticated systems, or simply because it is an easy way to pass risk down the contractual chain. Either way, they employ STCs. It is therefore very difficult to distinguish the different types of behaviour without digging a little deeper. Indeed, for STCs and their employers, the two types of objective may be combined. This fundamental conflict remains to be resolved, not least because of the difficulty of detection. It is a major source of the problems facing the construction industry. The problems range from lack of co-ordination, poor management, and low quality through to late payments, defective work and ultimately insolvency.

Cash flow management

A significant feature of construction firms is their pattern of cash flow. Low profit margins and low overheads are normal in the construction industry[9]. But as workload reduces and competition increases margins are reduced to maintain the flow of work. In stable and certain conditions this would be a reasonable strategy, but construction projects are characterised by uncertainty. Payments to contractors can be constrained by the client and consultant team: clients do not always pay promptly and consultants do not always certify fairly[10]. Even a good contractor has little power over such blatant abuses of contractual provisions but can mitigate the worst effects with strong and careful management. The contractor in turn controls payments to

---

[9] Government Statistical Service (1988) *Size analysis of UK businesses, 1988.* HMSO; London.

[10] *Michael Sallis & Co Ltd v E C A Calil and Others* [1988] 4 Const LJ 125; *Pacific Associates and Another v Baxter and Others* [1990] 1 QB 993; Chappell, D (1989) Is it worth suing and architect? *Building Today* 2 Feb, 24-25; Bingham, A (1992) A case to get you all steamed up. *Building* 3 Jul, **36**; *John Mowlem & Co plc v Eagle Star Insurance Co Ltd and Others* (1992) 33 Con LR 131.

subcontractors. The contractor can manipulate payments to subcontractors and suppliers to offset problems with receiving payment. However, this behaviour can increase the level of uncertainty because the contract structure permits unscrupulous management and aggressive financial practices. Of course, less able and less scrupulous contractors exacerbate such situations. The contractual issues arising from these problems are dealt with later in this paper. In this section the commercial problems are highlighted.

The effect of delayed payment

Contractors' cash flows are very sensitive. Although a contractor's mark-up is usually portrayed as a proportion of the contract sum, the true picture only emerges when it is viewed in the light of cash flow and use of capital. The simple fact is that contractors are paid periodically and pay for their supplies periodically. This means that if they receive payment before they incur liability to pay their own bills, they have no capital tied up in the project. On the other hand, delays in the receipt of their money change the picture completely, especially if they cannot delay the payments they must make for their supplies. Thus, if profit is related to the capital tied up in a project, it is clear that even a slight delay in receiving payment can turn a profit into a loss. The same argument applies all the way down the contractual chain to subcontractors and their suppliers.

When mark-up is related to the use of capital, two things are revealed; first, the potentially high profits for contractors (and the appeal of such lucrative business) and second, the vulnerability of this return. Every firm must avoid prolonged negative cash flows. The vagaries of bankers' lending policies and the manipulation of interest rates by governments add to the difficulties. Construction firms are particularly sensitive to disruptions in their cash flows, and therefore are more susceptible than most to changes in banking and government policies.

Payment

To reduce uncertainty it is vitally important that payments are made according to the agreement. Similarly it is important that the contract documents form an accurate record of the agreement. The results of the survey show that most payments are delayed. On average, STCs had to wait 11.5 days beyond the period stipulated in their contract, with only 15% being paid on time.

Reductions to payments

When margins are small it is important that completed work is valued fairly and paid in full so that the STC's cash flow is not threatened. However, for the reasons given above, certain practices are used by main contractors to mitigate their cash flow problems. These practices starve STCs of cash, thereby benefiting main contractors at STCs' expense. The survey showed that these practices are common. Although an STC may be spared from these practices on one site, their presence on another site may easily have a "domino effect", especially when the victim becomes insolvent.

Under-valuation

There are often disputes over the measurement and valuation of STCs' work, particularly where an item has not been specified in the bill of quantities. One reason for this may be a contractor's reluctance to commit expenditure on STCs' unspecified work before agreeing it with the employer. Another is that contractors like to retain a financial buffer. Under most procurement methods, contractual mechanisms are provided to protect STCs from the worst excesses of some contractors. Because of this, a chain of measurement and approval follows the chain of contracts providing many opportunities for negotiation and dispute.

Set-off

Set-off can also reduce payments to subcontractors. This covers contractors for the expense of employing a replacement subcontractor to finish off the work if necessary. The survey revealed a high incidence of spurious counter-claims aimed at retaining as much money as possible. It also showed that genuine claims were rarely pursued through the courts and were usually settled by negotiation.

Pay-when-paid

The interviews and the survey revealed that even without a pay-when-paid clause in the contract it is quite common for contractors to withhold payments to subcontractors until their own payment has been received. This is another technique by which contractors can protect their own cash flows.

Variations

Variations are often not valued until long after their issue, which has the effect of delaying or even avoiding payment for the work. Many variations occur as a natural part of the evolution of the design details. This is difficult to trace and sometimes can only be adequately evaluated at the end of the project. When left this late it inevitably results in claims, counter-claims and protracted arguments. Variations clearly attributable to the client are not as problematic.

Final accounts

Further problems arise from unreasonable delays to final account settlement. When work is inadequately specified, or subject to excessive variations, there is inevitably much re-measurement and negotiation to be done after completion. The interviews have confirmed what earlier research studies[11] have revealed: contractors, clients and consultants sometimes have little will to settle final accounts when they have got plenty of other, more lucrative work to distract them. Complex projects take even longer to settle. The final account process can to take up to ten years on major projects, being dependent upon the relationship between the client and design leader. The delay may be influenced by the implications for the design leader's PI insurance. Because of this, contractors and especially subcontractors must wait much longer than the payment period specified in their contracts. Only contractors and STCs with large cash reserves and/or healthy cash flows on other projects can sustain this level of disruption to their cash flow.

Background to procurement options

Much of the considerable experimentation with forms of procurement has been motivated by the desire to increase the interaction between the design and production processes[12]. The traditional lack of interaction between the design and production processes[13] has been exacerbated by the emergence of technologically sophisticated specialists whose work inherently

---

[11] Hughes, W P (1989) *The organisational analysis of building projects.* Unpublished PhD thesis, Faculty of the Built Environment, Liverpool John Moores University.

[12] Honeyman, S (Chmn) (1991) *Construction management forum: report and guidance.* Centre for Strategic Studies in Construction; Reading.

[13] Banwell, H (1964) *The placing and management of contracts for building and civil engineering works.* HMSO; London.

involves some kind of specialist design. The traditional general contracting approach to procurement was based upon the assumption that the contractor tendered on a complete design, and had no design responsibility. Since the specialist subcontractor had a design contribution, and since the project design team needed this design information early in the project, it was inevitable that the traditional approach would prove inadequate, encouraging the use of other approaches.

The following sections do not attempt to define the procurement methods as such, since that has already been done elsewhere[14]. Instead, they focus upon the role of STCs within each method, and serve as an introduction to the contractual problems facing STCs and those who wish to use them.

General contracting

The most common form of procurement is still general contracting[15], typified by contracts such as JCT 80 and ICE 6 in the UK. There have been many developments to these basic forms to attempt to integrate the STC's design into the project design team's effort, notably various methods of nominating subcontractors or suppliers. Without the formal contractual mechanisms of nomination, general contracting inhibits proper links between the STC and the design team, all communication and liability flowing through the main contractor who will have no design liability. This can produce vague and uncertain patterns of responsibility and liability and, consequently, much dispute.

Design and build

Design and build is an equally established procurement method, but one that originated with the express purpose of establishing

---

[14] Masterman, J (1992) *An introduction to building procurement systems.* Spon; London.
[15] Bound, C and Morrison, N (1991) Contracts in use. *Chartered Quantity Surveyor.* January, 9.

single-point responsibility. Under these forms of agreement, typified by JCT 81 and the ICE Design and Construct contract, the contractor takes on complete responsibility for design and construction. Traditionally, contractors have preferred to operate in this way when they have a special expertise. Designers have rarely advised its use except for projects that allow limited scope for innovation. More recently, however, with recessions biting into contractors' workloads, clients have invited contractors to tender based on this extended liability without having to pay a premium for the increased contractor's risk. Additionally, contractors have marketed their services more vigorously, and showed a willingness to take on such liability.

Design and build as a means of procurement places more responsibility and liability on to the contractor than traditional contracting. A major feature that separates design and build from general contracting is the lack of an independent certifying role for the lead designer. It is likely that a strict product liability would normally attach to a design and build contractor. The seemingly uncomplicated nature of the design-build arrangement may not be appropriate for all types of project. The signs are that its use is growing not because of its technical benefits, but because of the general shortage of work coupled with the possibility that many clients may have become disillusioned with other approaches. The presumed clarity of the arrangement appeals to clients, this in turn provides the marketing opportunity to contractors to promote its development as a procurement option. However, unless the choice of designer and specialists is left to the contractor, their involvement with the project can muddle its possible simplicity.

Management contracting

As the technological complexity of projects increased, STCs became more sophisticated and it became increasingly necessary to manage their contribution, both in terms of design before construction, and physically during construction. In addition, commercial exigencies during boom periods enabled STCs and contractors to charge significant risk premiums in their tenders; where the contractual or commercial risks were high, tenders were high. The logical conclusion to this process is that the lowest bid often came from the contractor who had failed to appreciate the risks involved in the project.

The relationship between risk and tender premiums was a critical factor in the emergence of management contracting. In a technologically sophisticated development project, the contractual risks for the contractor are so high as to lead to inflated tenders. The clients for such a project may be in a stronger position to be able to bear the risk, especially if the client is a property developer who builds frequently. It is a fundamental principle of risk apportionment that where a client builds frequently, and has large resources, the uncertainty associated with contractual risk is reduced. This is analogous to the reasoning behind the government's lack of fire insurance for their buildings. They have so many buildings that the mathematical probability of a fire occurring somewhere approaches certainty. Therefore it is meaningless to pay someone else to absorb it. This is why it is in the interests of clients to be able to choose contractual forms which reduce, or eliminate, contractual risk for the contractor.

The dual pressures of a keener need for co-ordination, and the client's desire to reduce the contractor's exposure to risk in the main contract led to the development of management contracting. A management contractor was one who would be appointed at an early stage of the process, effectively joining the design team, to advise on management and production issues, ensure timely involvement of the STCs' designers and manage the site processes.[16] The absence of contractual risk and the lack of any significant site work suited this more "professional" role. The way in which management contracting was subsequently used in practice often ignored the reasons for its development. The economic situation has worsened and there is little scope for contractors to allow for risk in their tenders for high-risk projects whilst they are hungry for work. In many cases, management contracts have been amended by clients whose advisers fail to appreciate the reasons for relieving the management contractor of contractual risk. This happens to such an extent that the disposition of risk becomes almost indistinguishable from that in general contracting and some of the worst features recur. These factors have combined to reduce the appeal and usage of management contracting.

Construction management

A more recent development is "construction management". The demand for this method of procurement has been fuelled by clients who have become frustrated with the difficulties of securing an efficient output from the industry. In construction management, the client employs all of the firms directly and provides the co-ordination and management through a consultant construction manager. In this way a cohesive design and construction organisation can be created, with the client absorbing the risks associated with co-ordination. A major feature of this approach is the way in which it elevates the status of the STCs to a major participating role which recognises their complete involvement with design and construction for their particular specialism. American and European practice is often cited as the origin of this technique, but in the context of the need to harness STCs without using vague or inappropriate

---

[16] CIRIA (1984) *A Client's Guide to Management Contracting in Building*. Special Publication No 33, Construction Industry Research and Information Association; London.

contract structures, it is clearly a logical development of UK practice. In any event, similar practices (for example, Separate Trades Contracting) have been common in the UK, especially in the North of England and Scotland, until relatively recently.[17] The major difference between UK and overseas practice is that UK designers seek to retain control over the final details, thereby maintaining the complexity of the information problem, but under the management of an integrated and controlled regime.

The integration of STC work into the procurement method

There are many other forms of procurement, most of which are simply variations on these basic themes. All of them have arisen as a response to deficiencies in traditional systems of contracting[18]. The traditional division between designers and constructors arises because of the professionalisation of design in the UK. This is the custom in which the design team documents the design on behalf of the client, who then seeks tenders from contractors. Whilst there is nothing wrong with this in principle, its effectiveness is compromised by two connected factors. First, designers in the UK are particularly skilled at developing their designs to a great level of detail and seek to retain control over its realisation in order to ensure that the details reflect the underlying design philosophy. This is not a problem on its own. Second, STCs have an increasingly important role to play in the design process, taking part in the development of detailed design information. This is the root of the difficult problem of enabling designers to have access to specialist knowledge without abrogating or compromising their own design responsibility and without breaching their liability. Whichever form of procurement is used, its choice should be influenced by the particular nature of specialist design for each project.

---

[17] Masterman, J (1992) *Op cit*
[18] Rougvie, A (1987) *Project evaluation and development.* Mitchell's; London.

Contractual issues in the procurement of STC work

The manner of the STC's engagement is largely dependent on the project procurement method. The survey, conducted as part of the study, found a wide diversity in the standard forms available, as well as the fact that they are rarely used unamended, these findings were comparable with a similar survey conducted by CASEC.[19] The amendments may often be trivial, but they frequently result in an employment regime not contemplated by those who originally drafted the standard forms. Amendments to contract clauses arise for two reasons. First, contractors and clients are sometimes guilty of arbitrarily shifting risk on to the weakest contracting party. Second, contract practice should be revised continuously in order that the documents reflect the rapidly changing technological and commercial nature of the construction industry. In other words, some of these changes to the standard forms are bad practice, others are good practice.

Payment

Management practices affecting cash flow have already been mentioned. Here we discuss typical contractual clauses and the amendments to standard-form contracts. The biggest single difficulty experienced by STCs is in getting paid for work they have done.[20] This has been confirmed by the survey, where payment was singled out as the most frequently amended standard contract clause. It is arguable whether a main contractor should have to pay a subcontractor before receiving any money. The main contractor's view was succinctly put by a management contractor who was interviewed:

---

[19] Greenwood, D J (1993) *Contractual arrangements and conditions of contract for the engagement of specialist engineering contractors for construction projects.* CASEC; London.

[20] Greenwood, D J (1993) *Op cit*

If we are getting a £100,000 fee on a £3,000,000 project, are we expected to act as a banker guaranteeing the subcontractors' payments? Pay-when-paid is not onerous on subcontractors from our point of view. Frequently, the clients use "shell" companies with no resources or capital. How can we guarantee payment from such organisations or assure the subcontractors of their financial stability? It is as simple as this: if we are not satisfied with a client's financial stability, then we don't build. Therefore, if we are going ahead, then clearly there is every likelihood of getting paid, and we don't need to be the client's guarantors.

Even though pay-when-paid clauses are not always present in subcontracts, the research has shown that contractors often operate in this way as a matter of policy. However, under these circumstances, the lack of payment to a subcontractor is a normal commercial risk taken by all businesses, and the lack of a pay-when-paid clause means that the subcontractor has a right to legal recourse.

Retention is a source of payment problems. The survey showed that the most onerous subcontracts tie the release of a subcontractor's retention to the employer's release to the main contractor. Retention can present a considerable obstacle to effective cash flow and across the UK industry large amounts of money are tied up in this way.[21] Most standard forms of subcontract allow the main contractor to deduct between 2.5% and 7.5% for prompt payment to the subcontractor. However, surveys have shown that it is normal practice for contractors to pay late and yet still withhold the discount[22]. This may be due to

---

[21] Middleboe, S (1992) Subbies plan mutual option for retention. *New Builder.* 1 Oct, 6.

[22] Confederation of Construction Specialists. (1992) *Corruption of the commercial process.* CCS; Aldershot: Barrick, A (1992) Payment scandal hits subbies and clients. *Building.*47 CCLVII, 20 Nov, 10.

the misleading practice in many standard forms of referring to this as a "cash discount", even though the intention is clearly to create a discount for promptness.

A further problem is set-off. Set-off is widely applied in commercial contracts and can be operated according to common law[23] or equitable principles.[24] It can to some extent be limited by use of express contractual provisions.[25] It was clear from the survey that spurious claims and contra-charges were causing severe disruptions to the cash flows of STCs. At a time of economic recession, these disruptions are sufficient to render firms insolvent. Clearly, many payment problems are a direct consequence of the recession. A main contractor who is suffering cash flow difficulties can temporarily counteract them by withholding payments to subcontractors. Thus, the solvency of one firm is assured only be jeopardising the solvency of another. The problem here is that, in practice, it is very difficult to distinguish these causes from other, more valid causes.

Nominated vs domestic subcontracts

The suggested reasons for the development of nomination were given earlier. It should, however, be remembered that the primary motivation behind nomination was to harness the skills of specialist subcontractors at an early stage. The most significant development has been the recent trend to oblige main contractors to use specialists without using the correct nomination procedures. Whilst wishing to retain control over the contractor's choice of subcontractor, many clients are specifying which firm must be used, but circumventing nomination by using the arrangements for domestic subcontractors. There is an

---

[23] *Mondel v. Steel* (1841) 8 M & W 858, 1 BLR 108.

[24] *Hanak v. Green* [1958] 2 QB 9

[25] *Gilbert Ash (Northern) Limited v. Modern Engineering (Bristol) Limited* [1974] AC 689; *NEI Thompson Limited v. Wimpey Construction UK Limited* (1987) 39 BLR 65.

inherent contradiction with this misuse of the contractual provisions. Typically, standard form contracts in the construction industry exclude design liability for the main contractor. Therefore, main contractors cannot assume design liability for their subcontractors, whether nominated or domestic. Contribution of design information from the STC may be essential to the success of the project, and the designer must have access to that design information. Under these circumstances, the main contractor cannot manage the design process. Therefore, this approach does not solve the fundamental design management problem and worse, it complicates and obscures the true picture.

Design warranties

> *Where specialist contractors undertake design, their appointment should be separated into two agreements and the design part should be subcontracted from the architect. The architect should have responsibility and should not be permitted to pass the buck. (Interview statement)*

Far from echoing the sentiment of the above statement, the interviews have generally shown that clients are seeking to resolve the design contract anomaly by seeking direct warranties between themselves and the STCs. These agreements are becoming very complex because each problem encountered by a client leads them to include an extra term in their own evolving standard version of a design warranty. Whilst it may well be good practice to keep these documents up to date by adding new terms, it is bad practice to retain permanently every item which has ever been considered necessary. This incremental approach to adding clauses leads to immensely complex documents covering every eventuality and requiring extensive legal advice merely to interpret them. This is exemplified by the following interview extract:

Only yesterday I received a contract document of 190 pages, much of it being the design warranty, for £10,000 of work.

This typifies many of the comments made by STCs who are increasingly worried about the complexity of the documents they are required to price. All of the specialist trade organisations interviewed said that they spend a considerable amount of their resources advising their members on the implications of one-off clauses.

Liability for late completion

A source of much discontent among subcontractors, especially small firms, is their liability in the event of a delay in completion. This discontent is expressed particularly in relation to those subcontracts that make provision for liquidated damages at a level equivalent to that in the main contract. To the subcontractors concerned, this seems unfair and excessive, since they are responsible only for part of the works, while the sums involved may well exceed their annual turnover. However, from the point of view of the main contractor, a subcontractor who delays the completion of the entire project can cause enormous losses; not only any liquidated damages for which the main contractor may become liable, but also the prolongation costs of both the main contractor and other subcontractors, to whom the main contractor is liable in turn. It would appear that if the main contractor is to be protected, every subcontractor must be potentially liable to this extent and further, then these liabilities should all be underwritten by bank or parent company guarantees.

However, the temptation to provide for such stringent liquidated damages in a subcontract is one which should be resisted. This is because a liquidated damages provision, in order avoid being struck down by the courts as a penalty, must represent a genuine pre-estimate of the employer's loss which is likely to result from delay. Pitching the liquidated damages provision in the subcontract at the same level as the main contract is almost bound to fail this test, since the main contractor will usually incur additional losses. Moreover, where a nominated subcontractor's delay is a ground on which the main contractor is entitled to an extension of time, LADs will not be payable under the main contract at all, so that the subcontract figure will again fail the "penalty" test.

Agreement to programmes

It is ironic that an STC acting as a subcontractor is usually required to perform exactly to the requirements of a main contractor, even though good and comprehensive programme information is rarely available from main contractors[26]. The requirements are particularly onerous when they are linked to the STC's liability for co-ordination with the work of other STCs. Together, these two factors impose an obligation upon STCs to adhere strictly to a poorly defined programme, and to integrate their own work with others.

Protection of work

A commonly found clause requires STCs to be responsible for the protection of their work from damage "howsoever caused". This can apply irrespective of whether the STC is still on site, and irrespective of the stage of the main contract. Clearly, this level of risk for the STC is difficult to price, and expensive (if not impossible) to control.

---

[26] Greenwood, D (1993) *Op cit*

Responding to amended contracts

The survey showed that over 80% of STCs qualify their bids, and the remainder add qualifications during the negotiations which lead to the signing of the contract. More than half seek legal advice and 31% consult their trade associations. The responses to the survey and information from the interviews indicate that the bidding process has two distinct stages. First, the STC prices the work to be done and negotiations firm this up. The second stage is the negotiations about the risk elements. STCs either seek to amend the clauses or negotiate a suitable payment for taking the risk. The success of STCs in dealing with risks in this way is dependent upon the relative economic power of the two negotiating parties, and the extent to which a main contractor needs to ensure consistency between subcontracts and the main contract terms (commonly referred to as ensuring that the contracts are "back-to-back").

The fact that so many STCs qualify their bids and enter into negotiations about contractual terms indicates that there are many firms who understand the significance of the risks that are being allocated to them. The survey and interviews revealed that a large proportion of contractors, consultants and subcontractors have little real understanding of the contracts they are using. The confusion, particularly with regard to standard forms of contract, is the source of many severe problems. It is important for all involved in the constructrion process to be more aware of the relationships between the business deals that they do and the contracts which purport to record their deals. While this problem is not unique to the construction industry,[27] the extent to which standard forms are relied upon in construction is particularly worrying.

---

[27] Beale H and Dugdale, A (1975) Contracts between businessmen: planning and the use of contractual remedies. *British Journal of Law and Society*, **2**, 45-60.

Good commercial and contracting practice

Many of the bad practices that have been exposed cause immense harm to the whole industry, not just to the immediate victims. For example, a main contractor who forces a subcontractor into liquidation faces problems of disruption. Someone in the team faces the need to find a replacement and the client will face escalating costs. The discussions in this paper point to some clear conclusions:

- Although nomination has declined, there remains the need to harness the skills of specialists at an early stage in the process.
- Subcontract problems are dominated by misunderstanding and vagueness.
- Procurement choice should be influenced by the nature of specialist input for the job.
- Contracts should be negotiated and explicit: they should accurately record the deal that was struck.
- These is generally insufficient awareness of the relationships between legal doctrines of contract and the negotiation of business deals.
- Onerous contract terms should be avoided because they beget qualified bids, extra negotiations and rancour
- Good cash flow management is essential to the success of contractors and subcontractors
- Project requirements should be clearly defined to reduce payment uncertainty
- Subcontractors' potential liabilities should be underwritten

In conclusion, perhaps commercial practice in the construction industry will only improve when all parties recognise that their long term interests will be served by respecting the commercial needs of their trading partners.

20. Shaping the Contract Package: A Radical Alternative

Richard Dobson

Synopsis

This paper presents the author's views on existing traditional forms of contract procurement. The author is critical of the JCT form, presenting suggestions for improvement. Examples of an alternative approach to contract procurement are given.

Introduction

I would perhaps have preferred to entitle my paper: "Re-shaping the contract package: Practical alternatives." I would like to start by commenting on how we, the construction industry, the largest industry in the country, satisfy the needs of our market. In simplistic terms our clients have 3 main criteria when selecting their product:

- Quality;
- Cost; and
- Programme

Depending on the type of client, the type of building required and the time available, the relative importance of these criteria vary. It is quite possible to envisage each one of these criteria being of first, second, third or even of equal importance to the others, depending on the particular circumstances. Having identified the

Client's priorities in these areas, I believe we, as an industry, should be able to deliver the goods with some degree of certainty. I believe we fail to do this too often to be acceptable to ourselves, let alone our clients. If we are to encourage our market to buy our product, then we must offer them greater security in their chosen specification of cost, quality and time.

Mankind has been building for as long as history itself. The building industry can perhaps even lay claim to being the second oldest profession! Yet still, we can not provide our customers with the buildings they want at the cost they bargained for and in the time available, with any real degree of certainty. We should ask ourselves, why is it that, at this stage in our development, the construction industry needs the contribution of a member of the government to remind us of the basic facts of life about what we should be producing and how we might go about achieving it? Like a breath of fresh air, Sir Michael has demonstrated our shortfalls and given us suggestions for improvement. May I suggest some further improvements worthy of consideration:

First, it is an unfortunate fact of life in our society that, however well considered a form of contract is, if the parties disagree to any substantial extent, then the judicial process can be used to 'steam-roller' any carefully constructed contract to the point of irrelevance. Using the most sophisticated of our standard forms of contract with all parties carrying out their duties within the terms of the contract, it is quite possible to fall into the litigation poker game of weighing up legal costs and likely judicial outcome against what you know to be the right and reasonable account. Never again do I want to have to advise a client that, even though he and his consultants have all carried out their responsibilities to the full, the financial stakes of proceeding into arbitration literally dwarf the amount of the original dispute to such an extent that a negotiated settlement is very often the only sensible option.

The simple arithmetic for the client is that in order to go into battle to defend a claim of, say, £1m, then ,depending on how successful you are in assessing the amount of a sealed offer, you have to be prepared to fork out for the legal costs of both sides, perhaps another £0.5m, together with the amount of the claim that might be adjudged to be due: cynically, this may be at least another £300,000. In other words, the client would have to risk losing almost the total amount of the claim made against him in order to defend his position as advised by his professionals. At this stage the contract might just as well be written on one side of A4 and it would probably better for all concerned if it were. The moral of this familiar story is quite clear to me: whatever the form of contract used, then it must do everything possible to minimise the risk of dispute. Sounds easy doesn't it? But isn't this made difficult by using our traditional contracts. Let me list the main sources of dispute in our contracts:

- Variations;
- Provisional and PC sums;
- Extensions of Time; and
- Claims for loss and expense.

I maintain that, in order to minimise the risk of entering the litigation poker game, we should take these out of our contracts altogether! Let me say that again in case anyone thought they must have misunderstood: I would advocate that Variations, Provisional and PC Sums, Extensions of Time and Claims for Loss and Expense have caused, and are still causing by far and away, the most difficulties and should be deleted from our contracts in their entirety. I feel the JCT over the years have been too accommodating to the apparent demands of its constituent bodies and have allowed these mechanisms into their contracts to the detriment of achieving the end product our clients really want. I feel that we have concentrated too much on our individual interests and too little on what our market really needs.We have ended up with a set of contracts which in my view encourage uncertainty, encourage confrontation, discourage teamwork and confuse lines of responsibility. These four monsters just mentioned are the main culprits.

Variations

Human nature is such that it is very difficult to make a final decision whilst there is still time to change your mind. It is a very rare human being indeed who can make a final decision - and stick to it - before the time available has elapsed. The allowance of variations in our contracts merely fosters this human weakness and is a positive incentive not to decide on issues of design and specification until the works are well under way. I cannot imagine any other commercial industry which would press the green button on the production line with so much uncertainty as to the final product. It would seem an impossible task to predetermine every last nut and bolt before you start on site. Every building is unique, how can we pre-design to that extent? This is generally correct, but given the right encouragement and framework, you would be amazed how much can be pre-defined and how little need be left for later. What greater incentive could there be for making your mind up pre-contract than the lack of a facility post-contract to make variations? The mind is concentrated wonderfully, but in most cases your first decision is the best decision in any event. If this seems a little unfair to the architects and engineers amongst you, let me offer you some help in this area.

We must collectively convince our clients to give us more time before we go into contract, and we must use that time more effectively. We are mostly to blame for irresponsibly rushing into contract because a client is anxious to see some action on site. We must educate our market to recognise the benefits of a longer pre-contract period with the resultant shorter and more certain construction period. We really should target the final completion more specifically, rather than rush to achieve the earliest start in the hope that the rest will follow on to suit. The other major assistance that would benefit hard-pressed decision makers is the contribution that the Contractor can make at the pre-contract stage. I feel that it is particularly bizarre that we consultants can very often beaver away at a scheme for many months, sometimes even years, and the poor old Contractor who has to actually build the thing, sometimes only gets a couple of weeks notice to start on site! Whether it be by negotiation or by a two-stage tender process or whatever, I am firmly of the view that there can be a tremendous benefit to the project generally if the Contractor is involved early on. Not only can the Contractor help the decision making process, suggest alternatives, steer us towards better buildability, but also help to prioritise our efforts at this critical stage.

The other major benefit of course is that the Contractor becomes "involved" with the project. He has helped to produce it, he assumes a greater responsibility to the end product and of course is much further up the learning curve by the time he starts on site. In a fixed sum contract without variations, it is the contractor who will be assessing and pricing the risk of those areas left undefined. It is his great incentive therefore to encourage those risk areas to be minimised and for the whole team to concentrate on and work like mad to reduce these risk areas to a minimum. I refer particularly to issues such as site and soil investigations, statutory requirements and charges, party wall issues, planning conditions and so on. Without variations, the Contractor can confidently plan his approach to the works and will have no-one to blame but himself if things do not go entirely to programme. If, during the contract, there emerges the unavoidable need to change something because of circumstances that no-one could have possibly foreseen, then there is nothing to prevent this "change" being incorporated by negotiation as a collateral contract appended to the original contract and a single price adjustment agreed between parties.

Provisional and PC Sums

Much the same principle applies to these two "nasties" as it does to Variations. How on earth can a Contractor sensibly assess his programme and site overheads requirements, with large chunks of Provisional Sums included in his contract? The arrangement within SMM 7 for "Defined" and "Undefined" Provisional Sums has helped the Contractor to a large extent, but left the poor old Client exposed to much greater risk. My suggestion is simple; do away with them altogether! They are, after all, only large areas of variation waiting to happen! PC Sums or even Named Sub-Contracts can be a great source of discontent. With the Contractor on board early on, it should however be possible to negotiate any preferred sub-contractors into the main contract package as mutually selected domestic sub-contractors with full responsibility for liaison, co-ordination and performance resting where it should be, with the Contractor. This leaves us with the two remaining major "bug-bears"of Extensions of Time and Claims for Loss and Expense.

Extensions of Time

I have to admit that I take a pretty hard line on the Extensions of Time issue. Let's have a look at the most common "Relevant Events" to see how many we can get rid of.

Exceptionally Adverse Weather

This is a much abused reason for delay as we all know. I would take it out of the contract altogether, and invite the Contractor to assess and take the risk on this one. Before you shout "unfair", let me ask you to consider who is the best equipped to judge how extremely bad weather could affect his building process? Who also is in the best position to minimise this risk by making the best of good weather conditions? It is, of course, the

Contractor. How many farmers do you know who enjoy the relief of an extension of time clause for bad weather? Let's encourage the Contractor to make hay whilst the sun shines, rather than preparing claims for a rainy day! At one stroke we can avoid those perennial arguments about what constitutes "exceptionally adverse weather" and what is simply normal British weather.

Insured Risks or Specified Perils

On the basis that the consequences of these risks are covered by insurance, then I would retain this clause, but I am close to being persuaded that it is more practical for the Contractor to take the risk of a time delay caused by, for example, fire, flood, lightning, storm, etc. After all, he is in the best position to take precautions to prevent damage to the Works caused by one of these eventualities.

Civil Commotion, Strikes etc

This is a risk that all of us in business take. Why should the Contractor get the benefit of a relief from this risk? I don't see why a Contractor cannot sensibly assess and accept this risk like the rest of us in business do.

Late Instructions

I would argue for this to be retained. The likelihood of late instructions is much less with this re-shaped contract as so much more has been defined pre-contract. I would also cynically observe that without the ability to euphemistically describe late instructions as exceptionally adverse weather (it often happens!), then the Contract Administrator may be encouraged to issue instructions more promptly!
Postponement by Employer would, however, remain as a valid reason for delay. But as far as variations and provisional sums

are concerned, if any "changes" are negotiated then these would become collateral contracts and the contract timetable could be amended by negotiation. As you can see, by taking a more pragmatic approach to variations and provisional sums, then one can begin to reap the benefit of much reduced areas of dispute on the question of delay and Extensions of Time.

Loss and Expense

This is the final demon which makes a vast contribution to the element of dispute in our building contracts. In a similar way to Extensions of Time, one can eliminate several main entitlements to Loss and Expense. As they no longer exist, there is no possibility of disruption caused by Variations,, Provisional and PC Sums, Nominated and Named Sub-Contractors and Approximate Quantities. All these hoary old chestnuts which occupy countless hours of speculation, debate and argument, simply disappear overnight. One is merely left with the possibility of eminently legitimate claims for Loss and Expense as a consequence of Delayed Possession, Late Instructions, Postponement of Work and not forgetting, of course, Failure to give Access to the Site. These are obvious candidates to keep and should not generate too much dispute even if called upon.

I believe that by taking these demons out of our contracts, we can reduce the risk of dispute enormously. You cut off those avenues of excuse that so often lead to ill will and distract the team from the common objective of achieving the right product at the right cost and at the right time. By bringing in the Contractor early, you are also building a more cohesive team with a clearer understanding of each other and the common objective of providing what the client has asked for. By agreeing a lump sum fixed price contract amount, with no facility for variations, you allow the Contractor the opportunity of carrying out his job properly and with a clarity of purpose. He is able to

plan his work and programme with confidence in the knowledge
that if things start to go wrong then there is no convenient escape
clause in the contract to dive down and he must make every
effort himself to correct the situation. All that energy that is
normally expended on fighting the negative issues, trying to
minimise one's individual responsibilities and preparing claims
and counter-claims can be diverted to the positive purpose of
collectively generating the end product that the Client has
ordered and has every right to expect, on time, to the right
standard and at the agreed cost.

Well, this sounds all rather wonderful doesn't it? But could it
really work? Is it just pie in the sky and wishful thinking? All I
can say to that is that we have tried it, on both new build and
refurbishment contracts and can report almost total success.
That's what makes me so enthusiastic about it and so committed
to helping others enjoy the same benefits. Let me give you, by
way of example, a short synopsis of our last completed project
carried out on this basis. It was a new build Housing Association
scheme in Camden. 27 Dwellings were constructed on an
extremely difficult site and the contract value was around
£1,800,000.

The Housing Association sheme in question was originally
tendered on a competitive basis using a Bill of Approximate
Quantities. The lowest Contractor was then invited to join the
team and over the next two months, we all worked together to
complete the design; eradicate wherever possible any areas of
risk; complete the remeasure of the completed design; agree all
sub-contract packages leaving no nasty Provisional Sums
remaining; and our efforts culminated in the negotiation of a
fixed sum contract price - the Contractor taking the risk on the
minimal amount of uncertainty left in the Works. You may be
interested to know that the fixed contract sum negotiated, was
£50,000 lower than the original tender figure, much to the
delight of the Client. In other words, by chiselling away at the

uncertain and undefined areas, by fully defining the building requirements and by negotiating a reasonable premium with the Contractor to take on board the remaining elements of risk, we were able to make a saving against the Provisional and Contingency Sums included in the original tender.

The contract was let on the basis of an amended JCT Intermediate Form of Contract with a fixed contract sum, no variations, no Provisional Sums and very limited grounds for Extension of Time and Loss and Expense. The Bill of Quantities, having served its purpose in arriving at the Contract Sum, was then discarded and did not form part of the contract documents. Why should it? After all, there were no variations to value. The 56 week contract was completed just two weeks late, and at the hand-over ceremony, we were able to present the Client with the signed final account in the amount of the contract sum! Not only that, but I have never before experienced such a united effort between Client, Contractor and Design Team throughout the contract period and the resultant award-winning scheme is of the highest quality.

Concluding Remarks

If I may conclude by summarising some of the main points that I have been trying to make. First, I believe we must do everything possible to avoid disputes, leading as they so often do, to the "litigation trap". Secondly, I believe our existing JCT contracts provide too many avenues for parties to avoid their responsibilities. They also in my view, give every imaginable opportunity for dispute. Why else do you think that a whole industry has developed in order to give advice and assistance in preparing and resisting contractual claims? Thirdly, I feel the main culprits in our contracts are Variations, Provisional and PC Sums, and the provisions for Extensions of Time and Loss and Expense. I believe it is possible to drastically reduce the scope for these monsters even if we can't eliminate them altogether.

There are huge benefits to be gained from introducing the Contractor to the team at the earliest possible opportunity. Get him involved and committed to the project long before he starts on site and you will be amazed how much better the whole process can work. Fourthly, negotiate a Fixed Contract Sum. Without the hindrance of variations or Provisional Sums, a Fixed Sum Contract is entirely feasible and gives the Contractor a clear view to plan and carry out the contract in his preferred manner. In this regime, allow the various disciplines to take responsibility for what they are good at. Give the Architect and Engineer the time and encouragement to fully design the project before commencement. Allow the Contractor to plan and build effectively with a clarity of purpose unfettered by variations and claims for Extensions of Time and Loss and Expense. Let the QS contribute to the process with good estimating and negotiating skills. Finally, concentrate above all on trying to achieve what the Client has commissioned you to achieve rather than be distracted by the temptation to fight and protect one's individual corner at the expense of the overall objective.

One final comment. These suggestions cut across an inherent in-built culture in our building industry and to make these changes requires stamina, perseverance, patience and, most of all, courage. It also requires the support and a reasonable level of competence from each and every member of the team. But what else do our Clients have the right to expect? I hope some of you will take up the challenge to make these changes in approach, to re-shape the traditional contracts and achieve the results that we have already enjoyed.

Select Bibliography

Ashley, D.B., Dunlop, J.R. & Parker, M.M., Impact of Risk Allocation and Equity in Construction Contracts, Source Document 44, University of Texas at Austin, March 1989.

Barber, J.N., Quality Managements in Construction - Contractual Aspects, SP84, CIRIA, London, 1992.

Connaughton, J.N., Value by Competition: A guide to the competitive procurement of consultancy services for construction, SP117, CIRIA, London, 1994.

Flanagan, R. & Norman, G., Risk Management and Construction, Blackwell Scientific Press, Oxford, 1993.

Hayes, R.W., Perry, J.G. & Thompson, P.A., Management Contracts, R100, CIRIA, London, 1983.

Latham, Sir Michael   Constructing the Team, Final Report of the Government/Industry Review of Procurement and Contractual Arrangements in the Construction Industry, July 1994, HMSO, London.

Perry, J.G. & Thompson, P.A., Target and Cost Reimbursable Contracts R85, CIRIA, London, 1982.

Pugh, C. & Day, M., Toxic Torts, Cameron May, London, 1992.

Skipp, B.O.(Ed), Risk and Reliability in Ground Engineering, Thomas Telford, London, 1993.

Sweet, J., Legal Aspects of Architecture, Engineering and the Construction Process, West Publishing Company, New York, 1985.

Thompson, P.A. & Perry, J.G. (Eds), Engineering Construction Risks. A guide to project risk analysis and risk analysis. Thomas Telford, London, 1992.

Uff, J. & Capper, P., Construction Contract Policy, Centre of Construction Law & Management, London, 1989.

Uff, J. & Clayton, C.R.I., Recommendations for the procurement of ground investigation, SP45, CIRIA, London, 1986.

Uff, J. & Lavers, A., Legal Obligations in Construction, Centre of Construction Law & Management, London, 1992.

Index